T0327504

Spatial Audio Processing

Spatial Audio Processing

MPEG Surround and Other Applications

Jeroen Breebaart
Philips Research, the Netherlands

Christof Faller
EPFL, Switzerland

John Wiley & Sons, Ltd

Other Wiley Editorial Offices

John Wiley & Sons Inc., 111 River Street, Hoboken, NJ 07030, USA

Jossey-Bass, 989 Market Street, San Francisco, CA 94103-1741, USA

Wiley-VCH Verlag GmbH, Boschstr. 12, D-69469 Weinheim, Germany

John Wiley & Sons Australia Ltd, 42 McDougall Street, Milton, Queensland 4064, Australia

John Wiley & Sons (Asia) Pte Ltd, 2 Clementi Loop #02-01, Jin Xing Distripark, Singapore 129809

John Wiley & Sons Canada Ltd, 6045 Freemont Blvd, Mississauga, Ontario, L5R 4J3, Canada

Wiley also publishes its books in a variety of electronic formats. Some content that appears
in print may not be available in electronic books.

Anniversary Logo Design: Richard J. Pacifico

British Library Cataloguing in Publication Data

A catalogue record for this book is available from the British Library

ISBN-13 978-0-470-03350-0

Typeset in 10/12 Times by Laserwords Private Limited, Chennai, India

Contents

Author Biographies

Jeroen Breebaart was born in the Netherlands in 1970. He studied biomedical engineering at the Technical University Eindhoven. He received his PhD degree in 2001 from the Institute for Perception Research (IPO) in the field of mathematical models of human spatial hearing. Currently, he is a senior scientist with Philips Research. His main fields of interest and expertise are spatial hearing, parametric stereo and multi-channel audio coding, automatic audio content analysis, and generic digital audio signal processing algorithms. He published several papers on binaural hearing, binaural modeling, and spatial audio coding. His work is incorporated in several international audio compression standards such as MPEG-4, 3GPP, and MPEG Surround.

Christof Faller received an MS (Ing) degree in electrical engineering from ETH Zurich, Switzerland, in 2000, and a PhD degree for his work on parametric multi-channel audio coding from EPFL, Switzerland, in 2004. From 2000 to 2004 he worked in the Speech and Acoustics Research Department at Bell Laboratories, Lucent Technologies and Agere Systems (a Lucent company), where he worked on audio coding for digital satellite radio, including parametric multi-channel audio coding. He is currently a part-time postdoctoral employee at EPFL. In 2006 he founded Illusonic LLC, an audio and acoustics research company. Dr Faller has won a number of awards for his contributions to spatial audio coding, MP3 Surround, and MPEG Surround. His main current research interests are spatial hearing and spatial sound capture, processing, and reproduction.

Foreword

It is a pleasure and great honor for me to contribute a foreword to this fascinating book on the state-of-the-art in stereo and multi-channel audio processing. Given my own research interest in binaural hearing, it is exciting to follow the detailed description of how scientific insights into human spatial perception in combination with digital signal processing techniques have enabled a major step forward in audio coding. I also feel honored to have a close relationship with the two authors. Both are young scientists with already an impressive output of scientific publications and also patents, and they both have made significant contributions to international standards. The book that they present here documents their deep insights in auditory perception and their ability to translate these into real-time algorithms for the digital representation of multi-channel sounds. I was lucky to follow many of the described developments from close by.

A remarkable aspect of the authors' careers is that both are or have been related to research environments with a long history in using perception insights to steer technological developments. Christof Faller was for many years affiliated with the Bell Laboratories of Lucent Technologies and later with Agere, a spin-off of Bell Laboratories and Lucent. The history of psychoacoustics at Bell Labs started around 1914, when Harvey Fletcher initiated a research program on speech and hearing with the clear goal to improve the design of telephone systems. Since then, many important contributions to audio and speech signal processing applications have come out of this laboratory. Jeroen Breebaart got his academic training at the former Institute for Perception Research (IPO), for many years a joint research institution of Philips Research and the Technische Universiteit Eindhoven. The transfer of his psychoacoustic modeling knowledge into algorithmic applications started after he had joined the Philips Research Laboratories in Eindhoven. Hearing research has been a topic at Philips since the 1930s when the technical possibility of stereophonic sound reproduction required a deeper understanding of spatial hearing. Among the many studies done in that period by, among others, K. de Boer were early applications of a dummy head to study and improve interpersonal communication and support hard-of-hearing subjects.

The book nicely demonstrates how close coding applications have come to present-day psychoacoustic research. When perceptual audio coders were first realized in the 1980s, the bit-rate reduction mainly exploited the concept of spectral masking. This concept was included in the encoder by so-called spectral spreading functions, an approximation which had been known in basic research for at least 30 years. Although using only a very crude description of perceptual processes, these early coders became essential in enabling the internet to be used for music distribution. Up to very recently, audio coding did not take much advantage of the redundancy *between* channels. The problem here is that, despite the *perceptual* similarity between different channels in a recording, the similarity in terms of interchannel correlation is often very low. Using a signal-based analysis thus does not give much room for redundancy removal. In order to capture and remove this redundancy, a time–frequency analysis and parametrization of the *perceptually relevant* spatial parameters is needed. The insights into these perceptual relations is of very recent origin, and the authors were able to apply them so quickly in audio processing because they had, in part, been involved in generating this perceptual knowledge.

Thus, the book is a remarkable document of what can be achieved by combining such complementary knowledge areas as psychoacoustics and digital signal processing. Given the economic and societal impact of audio compression, I hope that this example will help to attract future students to delve into this, certainly not simple, but highly rewarding research domain. Reading this book will certainly help readers to come to such a career choice.

Armin Kohlrausch

Research Fellow,
Philips Research, Eindhoven
Professor for Auditory and Multisensory Perception,
TU Eindhoven, the Netherlands

Eindhoven, May 2007

Preface

The physical and practical limitations that have to be considered in the field of audio engineering include limited directionality of microphones, limited number of audio channels, the need for backwards compatibility, storage and transmission channel constraints, loudspeaker positioning, and cost, which may dictate other restrictions.

There is a long history of interaction between the fields of audio engineering and psychoacoustics. In many cases, the reason for this interaction is to achieve a certain goal given the imposed limitations. For example, in the 1970s the first widely used cinema surround systems applied psychoacoustic knowledge to improve their perceptual performance. Specifically, due to technical limitations the method of representing multi-channel audio often caused a lot of crosstalk between various loudspeaker signals, resulting in a risk that the front dialogue was occasionally perceived from behind. To prevent this, delays were applied motivated by the psychoacoustic 'law of the first wavefront' (sound is often perceived from the direction from which the 'first wavefront' arrives and delayed reflections arriving from different directions are not perceived explicitly). Another example is that the spectral content of the center dialogue loudspeaker in cinemas is modified such that the dialogue is perceived from above the center loudspeaker, i.e. from the center of the screen as is desired.

A more recent example of how psychoacoustic knowledge benefits audio engineering is perceptual audio coding. Invented in the late 1980s at Bell Laboratories, perceptual audio coders reduce the precision of audio waveforms to the minimum such that the error is just not perceived. Due to the psychoacoustic phenomenon of (monaural) masking, i.e. that one sound can render other sounds inaudible, the precision of an audio signal can be reduced in a signal-adaptive manner with hardly any audible impairment. The most prominent example of such a perceptual audio coder is MP3.

Spatial audio coding and processing, the focus of this book, comprises processing of audio signals considering spatial aspects of sound in relation with the abilities and limitations of the human hearing system. As explained in this book, the amount of information required to represent stereo or multi-channel audio can be significantly reduced by considering how humans perceive the spatial aspect of sound. More generally, this book also gives many examples of audio signal processing, considering spatial hearing, for achieving desired results, such as binaural audio processing and two to N-channel audio upmix.

The authors would gratefully like to acknowledge the help, support, valuable insights, comments, suggestions and observations by the following people (in alphabetical order):

Frank Baumgarte, Frans de Bont, Bert den Brinker, Thomas Eisele, Jürgen Herre, Gerard Hotho, Armin Kohlrausch, Jeroen Koppens, Peter Kroon, Juha Merimaa, Francois Myburg, Fabian Nater, Werner Oomen, Mykola Ostrovskyy, Erik Schuijers, Michel van Loon, Leon van de Kerkhof, Steven van de Par, and Martin Vetterli. Furthermore, the authors would like to thank their colleagues from Agere Systems, Coding Technologies, Fraunhofer IIS and Philips for their support in developing and exploiting various audio coders based on the technology explained in this book.

Jeroen Breebaart

Christof Faller

May 2007

1

Introduction

1.1 The human auditory system

The human auditory system serves several important purposes in daily life. One of the most prominent features is to understand spoken words, which allows people to communicate in an efficient and interactive manner. In case of potential danger, the auditory system may provide means to detect dangerous events, such as an approaching car, at an early stage and react accordingly. In such cases, the great advantage of the auditory system compared with the visual system is that it allows us to monitor all directions simultaneously, including positions behind, above and below. In fact, besides a 360-degree 'view' in terms of both elevation and azimuth, the auditory system also provides an estimate of the distance of sound sources. This capability is remarkable, given the fact that humans have only two ears and yet are capable of analyzing an auditory scene in multiple dimensions: elevation, azimuth, and distance, while recognition of a sound source might be considered as a fourth dimension.

But besides being a necessary means for communication and to provide warning signals, the human hearing system also provides a lot of excitement and fun. Listening to music is a very common activity for relaxation and entertainment. Movies rely on a dedicated sound track to be exciting and thrilling. Computer games become more lifelike with the inclusion of dedicated sound tracks and effects.

In order to enjoy music or other audio material, a sound scene has to be recorded, processed, stored, transmitted, and reproduced by dedicated equipment and algorithms. During the last decade, the field of processing, storing, and transmitting audio has shifted from the traditional analog domain to the *digital* domain, where all information, such as audio and video material, is represented by series of bits. This shift in representation method has several advantages. It provides new methods and algorithms to process audio. Furthermore, for many applications, it can provide higher quality than traditional analog systems. Moreover, the quality of the material does not degrade over time, nor does making copies have any negative influence on the quality. And finally, it allows for a more compact representation in terms of information quantity, which makes transmission and storage more efficient and cheaper, and allows devices for storage and reception to

Spatial Audio Processing: MPEG Surround and Other Applications   Jeroen Breebaart and Christof Faller
© 2007 John Wiley & Sons, Ltd

be of very small form factor, such as CDs, mobile phones and portable music players (e.g. MP3 music players).

1.2 Spatial audio reproduction

One area where audio systems have recently gained the potential of delivering higher quality is their *spatial* realism. By increasing the number of audio channels from two (stereophonic reproduction) to five or six (as in many home cinema setups), the spatial properties of a sound scene can be captured and reproduced more accurately. Initially, multi-channel signal representations were almost the exclusive domain of cinemas, but the advent of DVDs and SACDs have made multi-channel audio available in living rooms as well. Interestingly, although multi-channel audio has now been widely adopted on such storage media, broadcast systems for radio and television are still predominantly operating in stereo. The fact that broadcast chains still operate in the two-channel domain has several reasons. One important aspect is that potential 'upgrades' of broadcast systems to multi-channel audio should ensure backward compatibility with existing devices that expect (and are often limited to) stereo content only. Secondly, an increase in the number of audio channels from two to five will result in an increase in the amount of information that has to be transmitted by a factor of about 2.5. In many cases, this increase is undesirable or in some cases simply unavailable. With the technology that is currently being used in broadcast environments it is very difficult to overcome these two major limitations.

But besides the home cinema, high-quality multi-channel audio has made its way to mobile applications as well. Music, movie material, and television broadcasts are received, stored, and reproduced by mobile phones or mobile audio/video players. On such devices, an upgrade from stereo to multi-channel audio faces two additional challenges on top of those mentioned above. The first is that the audio content is often reproduced over headphones, making multi-channel reproduction more cumbersome. Secondly, these devices are often operating on batteries. Decoding and reproduction of five audio channels requires more processing and hence battery power than two audio channels, which has a negative effect on a very important aspect of virtually all mobile devices: their battery life. Furthermore, especially in the field of mobile communication, every transmitted bit has a relatively high price tag and hence high efficiency of the applied compression algorithm is a must.

1.3 Spatial audio coding

Thus, the trend towards high-quality, multi-channel audio for solid-state and mobile applications imposes several challenges on audio compression algorithms. New developments in this field should aim at unsurpassed compression efficiency, backward compatibility with existing systems, have a low complexity, and preferably support additional capabilities to optimize playback on mobile devices. To meet these challenges, the field of spatial audio coding has developed rapidly during the last 5 years. Spatial audio coding (SAC), also referred to as binaural cue coding (BCC), breaks with the traditional view that the amount of information that has to be transmitted grows linearly with the number of audio channels. Instead, spatial audio coders, or BCC coders, represent two or more audio channels by a certain *down-mix* of these audio channels, accompanied by additional information (spatial parameters or binaural cues) that describe the loss of spatial information caused by the down-mix process.

Conventional coders are based on waveform representations attempting to minimize the error induced by the lossy coding process using a certain (perceptual) error measure. Such perceptual audio coders, for example MP3, weight the error such that it is largely masked, i.e. not audible. In technical terms, it is said that 'perceptual irrelevancies' present in the audio signals are exploited to reduce the amount of information. The errors that are introduced result from *removal* of those signal components that are perceptually irrelevant.

Spatial audio coding, on the other hand, represents a multi-channel audio signal as a down-mix (which is coded with a conventional audio coder) and the before mentioned spatial parameters. For decoding, the down-mix is 'expanded' to the original number of audio channels by restoring the inter-channel cues which are relevant for the auditory system to perceive the correct auditory spatial image. Thus, instead of achieving compression gain by removal of irrelevant information, spatial audio coding employs *modeling* of perceptually *relevant* information only. As a result, the bitrate is significantly lower than that of conventional audio coders because the spatial parameters contain much less information than the (compressed) waveforms of the original audio channels. As will also be explained in this book, the representation of a multi-channel audio signal as a down-mix plus spatial parameters not only provides a significant compression gain, it also enables new functionality such as efficient binaural rendering, re-rendering of multi-channel signals on different reproduction systems, forward and backward format conversion, and may provide means for interactivity, where end-users can modify various properties of individual objects within a single audio stream.

1.4 Book outline

Briefly summarized, the contents of the chapters are as follows:

Chapter 2 provides an overview of common audio reproduction, processing, and compression techniques. This includes discussion of various loudspeaker and headphone audio playback techniques, conventional audio coding, and matrix surround.

Chapter 3 reviews the literature on important aspects of the human spatial hearing system. The focus is on the known limitations of the hearing system to perceive and detect spatial characteristics. These limitations form the fundamental basis of spatial audio coding and processing techniques.

Chapter 4 explains the basic concepts of spatial audio coding, and describes the interchannel parameters that are extracted, the required signal decompositions, and the spatial reconstruction process.

Chapter 5 describes the structure of the MPEG 'enhanced aacPlus' codec and how spatial audio coding technology is embedded in this stereo coder.

Chapter 6 describes the structure of MPEG Surround, a multi-channel audio codec that was finalized very recently. Virtually all components of MPEG Surround are based on spatial audio coding technology and insights. The most important concepts and processing stages of this standard is be outlined.

Chapter 7 describes the process of generating a virtual sound source (for headphone playback) by applying spatial audio coding concepts.

Chapter 8 expands the spatial audio coding approach to complex auditory scenes and describes how parameter-based virtual sound source generation processes are incorporated in the MPEG Surround standard.

Chapter 9 reviews methods to incorporate user interactivity and flexibility in terms of spatial rendering and mixing. By applying parameterization on individual objects rather than individual channels, several modifications to the auditory scene can be applied at the audio decoder side, such as re-panning, level adjustments, equalization or effects processing of individual objects present within a down-mix.

Chapter 10 describes algorithms to optimize the reproduction of stereo audio on different reproduction systems than the audio material was designed for, such as 5.1 home cinema setups, or wavefield synthesis systems.

2

Background

2.1 Introduction

Spatial audio processing and coding are not topics that can be treated in an isolated fashion. Spatial audio signals have properties which are related to the specific audio playback system over which the signals are intended to be reproduced. Further, specific properties also result from the microphone setup used if spatial audio signals are directly recorded. In addition to discussing spatial audio playback systems, recording is very briefly discussed. Additionally, conventional audio coding and matrix surround are reviewed.

2.2 Spatial audio playback systems

There has been an ongoing debate about the aesthetic aim of recording and reproducing sound. In recording of classical music or other events implying a natural environment, the goal of recording and reproduction can be to re-create as realistically as possible the illusion of 'being there' live. ('Being there' refers to the recreation of the sound scene at the place and time of the performance. The term 'there and then' is often used to describe the same concept, in contrast to 'here and now', which describes the sound scene at the place and time during playback.) In many other cases, such as movie sound tracks and pop music, sound is an entirely artificial creation and so is the corresponding spatial illusion, which is designed by the recording engineer. In such a case, the goal of recording and reproduction can be to create the illusion of the event 'being here', i.e. the event being in the room where playback takes place.

In any case, the requirement of a spatial audio playback system is to reproduce sound perceived as realistically as possible, either as 'being there' or 'being here'. Note that in 'being there' one would like to create the spatial impression of the concert hall 'there', whereas in 'being here' the acoustical properties of the playback room 'here' are to play a more important role. But these aesthetic issues are to be addressed by the performing artists and recording engineers, given the limits of a specific target spatial audio playback system. In the following, we describe three of the most commonly used consumer spatial audio playback systems: stereo loudspeaker playback, headphone playback, and multi-channel surround loudspeaker playback. A relation to Section 3 is established by linking spatial hearing phenomena to the described playback systems. A more thorough overview, covering the history and a wide range of playback systems not discussed here can be found elsewhere [250] and [227].

2.2.1 Stereo audio loudspeaker playback

The most commonly used consumer playback system for spatial audio is the stereo loudspeaker setup as shown in Figure 2.1(a). Two loudspeakers are placed in front on the left and right sides of the listener. Usually, these loudspeakers are placed on a circle at angles $-30°$ and $30°$. The width of the auditory spatial image that is perceived when listening to such a stereo playback system is limited approximately to the area between and behind the two loudspeakers.

Stereo loudspeaker playback depends on the perceptual phenomenon of summing local-ization, as will be described in Chapter 3, Section 3.3.4, an auditory event can be made to appear anywhere between a loudspeaker pair in front of a listener by controlling the *inter-channel time difference* (ICTD) and/or *inter-channel level difference* ICLD. It was Blumlein [28] who recognized the power of this principle and filed his now-famous patent. Blumlein showed that when only introducing amplitude differences (ICLD) between a loudspeaker pair, it would be possible to create phase differences between the ears (interaural time difference, or ITD) similar to those occurring in natural listening. He proposed a number of methods for pickup of sound, leading to the now common technique of *coincident-pair microphones*.

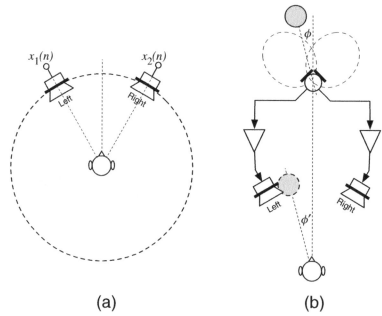

(a)   (b)

Figure 2.1   (a) Standard stereo loudspeaker setup; (b) coincident-pair microphone pickup and playback of the resulting stereo signal.

However, Blumlein's work remained unimplemented in commercial products for dozens of years. It was not until the late 1950s when stereo vinyl discs became available. These applied methods for cutting two-channel stereo signals onto a single disc similar to a technique already proposed in Blumlein's patent.

Capturing natural spatial sound

Figure 2.1(b) illustrates sound pickup and playback with coincident-pair microphones. Two directional microphones at the same location are orientated such that one microphone is headed more to the left and the other more to the right. Since, ideally, both microphones are at the same location, there is no phase difference (ICTD) between their signals. But due to their directionality there is an intensity difference (ICLD). For example, sources located on the left side result in a stronger signal in the microphone heading towards the left side than in the microphone heading towards the right side. In other words, the ICLD between the two microphone signals is a function of the source angle ϕ. When these microphone signals are amplified and played back over a loudspeaker pair, an auditory event will appear at an angle ϕ' which is related to the original source angle ϕ, as illustrated in Figure 2.1(b). If the recording system parameters are properly chosen one can achieve $\phi \approx \phi'$. When there are multiple concurrently active sources to be recorded (e.g. musical instruments playing together) the recording and playback principle already mentioned is also applicable and usually results in multiple auditory events, one for each instrument. More on this subject will be outlined in Section 3.3.5.

Coincident-pair microphones are a commonly used technique for stereo sound pickup. But there are a number of other popular microphone techniques. As mentioned, coincident-pair microphones ideally result in a signal pair without phase differences (ICTD = 0). This has the advantage that the resulting signal pair is 'mono compatible', i.e. when the signal pair is summed to a single mono signal, no problems will occur due to a comb-filter effect (cancellation and amplification of signal components which are out-of-phase and in-phase, respectively).

Early spatial audio playback experiments based on 'spaced microphone configurations' were carried out at Bell Laboratories [245]. In spaced microphone configurations the different microphones are located at different locations. Therefore, such techniques will result not only in ICLD cues, but also in ICTD cues. When the goal is to retain mono compatibility special care has to be taken when mixing the microphone obtained signals to the final stereo mix. It goes beyond the scope of this overview to describe such other microphone techniques in more detail. More on this topic can be found in [250] and [227].

Artificial generation of spatial sound

Artificial auditory spatial images for stereo loudspeaker playback systems can be generated by mixing a number of separately available source signals (e.g. multitrack recording). In practice, mostly ICLD are used for mixing of sources in this way, denoted *amplitude panning*. The concept of amplitude panning is visualized in Figure 2.2. One sound source $s(n)$ is reproduced using two loudspeakers with signal scale factors a_1 and a_2. The perceived direction of an auditory event appearing when amplitude panning is applied follows

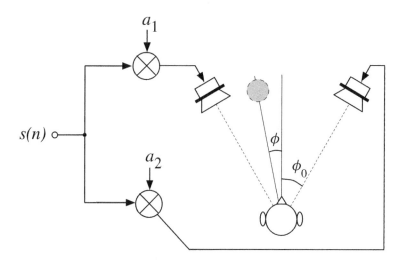

Figure 2.2 Definitions of scale factors and angles for the stereophonic law of sines (2.1).

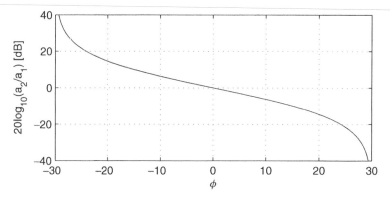

Figure 2.3 The relation between auditory event angle ϕ and ICLD, i.e. $20\log_{10}(a_2/a_1)$, for the stereophonic law of sines.

approximately the stereophonic law of sines derived by Bauer [11],

$$\frac{\sin\phi}{\sin\phi_0} = \frac{a_1 - a_2}{a_1 + a_2} \tag{2.1}$$

where $0° \le \phi_0 \le 90°$ is the angle between the forward axis and the two loudspeakers, ϕ is the corresponding angle of the auditory event, and a_1 and a_2 are scale factors determining ICLD. The relation between ICLD, i.e. $20\log_{10}(a_2/a_1)$, and ϕ is shown in Figure 2.3 for a standard stereo listening setup with $\phi_0 = 30°$.

Bennett *et al.* [15] derived a panning law considering an improved head model compared to the stereophonic law of sines. The result was a 'stereophonic law of tangents' which is similar to another earlier proposed law by Bernfeld [18], but for different listening conditions. Amplitude panning and auditory event direction perception is discussed in more detail in [210]. Note that all the mentioned panning laws are only a crude approximation since the perceived auditory event direction ϕ also depends on signal properties such as frequency and signal bandwidth.

A second method to reproduce a sound source at a desired position is referred to as *delay panning*. The implementation of delay or ICTD panning in analog mixing consoles would have been much more difficult than implementing amplitude panning. This was surely one reason why ICTD panning was hardly used. But even today, when implementation of ICTD panning would be simple in the digital domain, ICTD panning is not commonly used. This may be due to the fact that ICLD are somewhat more robust than ICTD when a listener is not exactly in the *sweet spot* (optimal listening position). ICLD may be perceived as being more robust because amplitude panning with large-magnitude ICLD results in auditory events at the loudspeaker locations by means of only giving signal to one loudspeaker. In such a case, an auditory event is perceived at the loudspeaker location even in cases when the listener is not in the sweet spot. This is one reason why in movie theaters usually a center loudspeaker is used, i.e. to have auditory events associated with dialogue at the center of the screen for all movie viewers. For pure ICTD panning signal is always given with the same level to more than one loudspeaker and a situation with stable auditory events (i.e. only signal from one loudspeaker) does not occur.

In addition to panning, artificial reverberation may be added to the stereo signal for mimicking the spatial impression of a certain room or hall. Other signal modifications may be carried out for controlling other attributes of auditory events and the auditory spatial image.

2.2.2 Headphone audio playback

Headphone stereo audio playback

Stereo audio signals are mostly produced in an optimized way for loudspeaker playback, as described in the previous section. This is reflected by the fact that during the production process the signals are usually monitored with loudspeakers by the recording engineer. The mixing parameters ICLD and ICTD result in relatively similar phenomena with respect to localization and lateralization of auditory events when a signal is presented over loudspeakers or headphones, respectively. Thus, one single stereo signal can be used for either loudspeaker or headphone playback. A major difference is that headphone listening with such stereo signals is limited to in-head localization, i.e. the width of the auditory spatial image is limited to being inside of the head of the listener as will be described in Chapter 3.

Headphone binaural audio playback

For regular audio playback (headphones, loudspeakers) the resulting *interaural time difference* (ITD) and *interaural level difference* (ILD) cues (i.e., the spatial cues at the level of the eardrums) only crudely approximate the cues evoked by sources that are physically

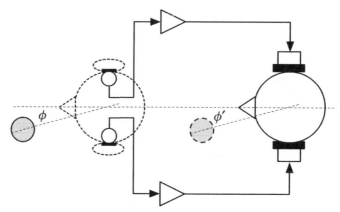

Figure 2.4 A binaural recording is a two-channel audio signal recorded with microphones close to the ear entrances of a listener or dummy head (left). When these signals are played back with headphones (right), a realistic three-dimensional auditory spatial image is reproduced mimicking the natural spatial image that occurred during recording.

placed at the auditory event positions. Furthermore, cues related to other attributes of the auditory events and auditory spatial image are also not entirely realistic, but determined largely by the recording engineer as a function of microphone setup parameters, mixing techniques, and sound effects processing.

Binaural audio playback aims at presenting a listener with the same signals at the ear entrances as the listener would receive if he were at the original event to be reproduced. Thus, all the signal cues related to perception of the sound are realistic, enabling a three-dimensional sound experience. Note that in this case the width of the auditory spatial image is not limited to inside the head (this is often called *externalization*).

Figure 2.4 illustrates a system for binaural recording and binaural headphone playback. During a performance, two microphones are placed at or near the ear entrances of a listener or a dummy head and the respective signals are recorded. If these signals are played back with binaural headphones, a listener will experience an auditory spatial image very similar to the image he would perceive if he would be present at the original performance. If the binaural recording and playback are carried out with a single person, the auditory spatial image experienced during playback is very realistic. However, if a different person (or dummy head) is chosen for the binaural recording the sound experience of the listener is often limited (front/back confusion, limited externalization). Front/back confusion can be avoided by modifying the signals as a function of a listener's head movements [30].

The dependence on the individual listener is one reason why binaural recordings are commercially hardly used. Another reason is that binaural recordings do not sound very good when played over a stereo loudspeaker setup. Several approaches have been proposed for playing back binaural recordings over two loudspeakers. Crosstalk cancellation techniques [5, 10, 130] pre-process the loudspeaker signals such that the signals at the ear entrances approximate the binaural recording signals. Disadvantages of this approach are that it only works effectively when a listener's head is located exactly in the sweet spot, and its performance at higher frequencies is limited. A technique where a binaural recording is post-processed with a filter with the goal of being comparable in quality to conventional stereo recordings for loudspeaker playback was proposed in [252–254]. This can be viewed as post-processing binaural recordings so as to mimic the properties of a good stereo microphone configuration. The idea is to store the signals obtained as conventional stereo signals. These signal would play back in good quality on standard stereo loudspeaker setups and when a device is intended for binaural playback it would incorporate a filter which would undo the post-processing that was applied prior to storage of the signal. The result would be a signal similar to the original binaural recording.

Binaural recordings can also be created by artificially mixing a number of audio signals. Each source signal is filtered with head related transfer functions (HRTFs) or binaural room impulse responses (BRIRs) corresponding to the desired location of its corresponding auditory event. The resulting signal pairs are added resulting in one signal pair. More detailed information on the synthesis of virtual sound sources and corresponding binaural cues is given in Chapter 6.

2.2.3 Multi-channel audio playback

Five-to-one (5.1) surround

Only in recent years have multi-channel loudspeaker playback systems become widely used in the consumer domain. Such systems are mostly installed as 'home theater systems' for playing back audio for movies. This is partially due to the popularity of the digital versatile disc (DVD), which usually stores five or six discrete audio channels designated for such home theater system audio playback. Figure 2.5 illustrates the standard loudspeaker setup for such a system [150], denoted *5.1 surround*. For backwards

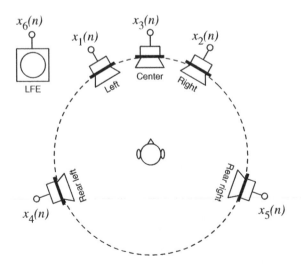

Figure 2.5 Standard 5.1 surround loudspeaker setup with a low-frequency effects (LFE) channel.

compatibility to stereo in terms of loudspeaker positions, in the front, two loudspeakers are located at angles $-30°$ and $30°$. Additionally, there is a center loudspeaker at $0°$, providing a more stable center of the auditory spatial image when listeners are not exactly in the sweet spot. The two rear loudspeakers, located at $-110°$ and $110°$, are intended to provide the important lateral signal components related to spatial impression. There is one additional channel (the .1 in 5.1) intended for *low frequency effects* (LFE). The LFE channel has only a bandwidth up to 120 Hz and is for effects for which the other loudspeakers can not provide enough low-frequency sound pressure, e.g. explosion sounds in movies.

The angle between the two rear loudspeakers is so large ($140°$) that amplitude panning between those loudspeakers is problematic. Similarly, it is problematic to apply amplitude panning between the front and rear loudspeakers. Thus, the standard 5.1 system is not optimized for providing a $360°$ general auditory spatial image, but for providing a solid frontal auditory spatial image with lateral sound from the sides for spatial impression. In terms of $360°$ rendering, it is never problematic to place an auditory event at the location of a loudspeaker. In this sense, also the 5.1 system provides some good possibilities for auditory events appearing either at rear left or rear right. With only five main loudspeakers, a system has to be a compromise, as reflected by the 5.1 system.

Capturing and generating sound for 5.1 systems

Different microphone configurations and mixing techniques have been proposed for generating sound for 5.1 systems, see e.g. [227]. Alternatively, techniques applied for recording or mixing two-channel stereo can be applied to a specific channel pair of the five main loudspeakers of a 5.1 setup. For example, for obtaining an auditory event from a specific direction, the loudspeaker pair enclosing the desired direction is selected and the corresponding signals are recorded or generated similarly as for the stereo case (resulting in auditory events between the two selected loudspeakers). Vector base amplitude panning (VBAP) [211, 212], when applied to two-dimensional loudspeaker setups such as 5.1, applies the principle in terms of amplitude panning (ICLD). But also other techniques have been proposed, feeding signals to more than two loudspeakers simultaneously, e.g. three loudspeakers [93].

2.3 Audio coding

Generally speaking, audio coding is a process for changing the representation of an audio signal to make it more suitable for transmission or storage. Although high-capacity channels, networks, and storage systems have become more easily accessible, audio coding has retained its importance. Motivations for reducing the bitrate necessary for representing audio signals are the need to minimize transmission costs or to provide cost-efficient storage, the demand to transmit over channels with limited capacity such as mobile radio channels, and to support variable-rate coding in packet oriented networks. In this section, representations and coding techniques which are of relevance to spatial audio are reviewed.

2.3.1 Audio signal representation

Audio signals are usually available as discrete time sampled signals. For example, a *compact disc* (CD) stores stereo audio signals as two separate audio channels each sampled with a sampling frequency of 44.1 kHz. Each sample is represented as a 16-bit signed integer value, resulting in a bitrate of $2 \times 44.1 \times 16 = 1411$ kb/s. For multi-channel audio signals or signals sampled at a higher sampling frequency the bitrate scales accordingly with the number of channels, bit depth, and sampling frequency.

2.3.2 Lossless audio coding

Ideally, an audio coder reduces the bitrate without degrading the quality of the audio signals. So-called lossless audio coders [51, 55, 95, 213, 223] can achieve this by reducing the bitrate while being able to perfectly reconstruct the original audio signal. Bitrate reduction in this case is possible by exploring redundancy present in audio signals, e.g. by applying prediction over time or a time–frequency transform, and controlling the coding process such that during decoding the values are rounded to the same integer sample values as the original samples. For typical CD audio signals, lossless audio coders reduce the bitrate by about a factor of 2. For higher sampling rates the higher degree of redundancy in the corresponding samples result in higher compression ratios.

Figure 2.6 illustrates a lossless audio coder based on prediction. It operates as follows:

Encoder: a linear predictor is used to predict the sample $x(n)$ as a function of the past samples $x(n-1), x(n-2), \ldots, x(n-M)$,

$$\hat{x}(n) = \sum_{i=1}^{M} h(i-1)x(n-1) \qquad (2.2)$$

where M is the predictor order. Note that for the prediction quantized prediction parameters are used in order that the same prediction can be carried out also in the decoder. The prediction error, $e(n) = x(n) - \hat{x}(n)$, is computed and lossless encoded together with the prediction parameters.

Decoder: the prediction error $e(n)$ and the prediction parameters $h(n)$ are decoded. The same predictor as in the encoder is used to compute $\hat{x}(n)$ given the previous samples (2.2), The estimated sample, $\hat{x}(n)$, is corrected with the prediction error, resulting in the decoder output $x(n)$.

Alternatively, lossless audio coders can be implemented using an integer time/frequency transform. Figure 2.7 illustrates a lossless audio encoder and decoder based on an integer filterbank, e.g. *integer modified discrete cosine transform* (intMDCT) [92]. An integer filterbank takes a time discrete signal with integer amplitude and transforms it to the frequency domain which is also represented as integer numbers. The encoder applies

Encoder:

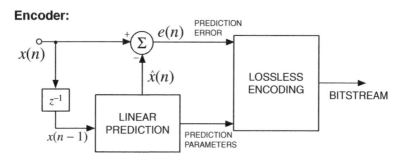

Decoder:

Figure 2.6 Generic prediction-based lossless audio encoder and decoder. Encoder: a linear predictor is used to predict the input integer samples. The prediction error is computed and coded together with the prediction parameters. Decoder: the prediction error and prediction parameters are decoded. The current prediction is corrected with the prediction error.

Encoder:

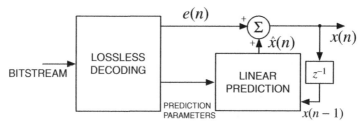

Decoder:

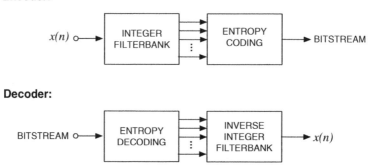

Figure 2.7 Generic lossless subband audio encoder and decoder. Encoder: the input signal is decomposed into a number of sub-bands by means of a perfect reconstruction integer filterbank. The integer sub-band signals are entropy coded and transmitted as bitstream to the decoder. Decoder: the coded sub-band signals are decoded and the inverse integer filterbank is applied to recover the audio signal.

an entropy coder, e.g. a Huffman coder, to the subband signals and transmits the coded subband signals as bitstream to the decoder. The decoder decodes the bitstream to recover the integer subband signals and applies the inverse filterbank to recover the original audio signal.

2.3.3 Perceptual audio coding

Most audio coders are 'lossy' coders not aiming at reconstructing an audio signal perfectly. The primary motivation for using lossy audio coders is to achieve a higher compression ratio. A *perceptual audio coder* is an audio coder which incorporates a *receiver model*, i.e. it considers the properties of the human auditory system. Specifically, the concept of auditory *masking* is employed. Masking refers to the fact that one signal that is clearly audible in isolation (the 'maskee') can become inaudible when presented simultaneously with another (often louder) signal (the 'masker'). The amount of masking depends strongly on the temporal and spectral content of both masker and maskee; in principle both signals should be both temporally and spectrally close for effective masking. In audio coders, the audio signal itself is the masker, while the quantization noise that is introduced due to data reduction is the maskee that should be kept inaudible. Hence audio coders employ a model to compute the *masked threshold* [293]. The masked threshold specifies as a function of time and frequency a signal level below which a maskee is not perceptible, i.e. the maximum level the quantization noise can have such that it is masked by the audio signal to be coded. By controlling the quantization noise as a function of time and frequency such that it is below the masked threshold, the bitrate can be reduced without degrading the perceived quality of the audio signal.

A generic perceptual audio coder, operating in a subband (or transform) domain, is shown in Figure 2.8. The encoding and decoding process are as follows.

Encoder: the input signal is decomposed into a number of sub-bands. Most audio coders use a *modified discrete cosine transform* (MDCT) [188] as time–frequency representation. A perceptual model computes the masked threshold as a function of time and frequency. Each sub-band signal is quantized and coded. The quantization error of each sub-band signal is controlled such that it is below the computed masked threshold. The bitstream not only contains the entropy coded quantizer indices of the sub-band signals, but also the quantizer step sizes as determined by the masked threshold (denoted 'scale factors').

Decoder: the entropy coded sub-band quantizer indices and quantization step sizes ('scale factors') are decoded. The inverse quantizer is applied prior to the inverse filterbank for recovering the audio signal.

This coding principle was introduced for speech coding by Zelinski and Noll [287]. Brandenburg *et al.* [41] applied this technique to wideband audio signals. Recent perceptual audio coders can be divided into two categories: so-called transform coders and sub-band coders. The first category employs a transform (such as an MDCT) on subsequent (often overlapping) segments to achieve control over separate frequency components. The second category is based on (critically sampled) filterbanks.

When more than one audio channel needs to be encoded, redundancy between the audio channels may be exploited for reducing the bitrate. *Mid/side* (M/S) coding [156] reduces the redundancy between a correlated channel pair by transforming it to a sum/difference channel pair prior to quantization and coding. The masked threshold depends on the interaural signal properties of the signal (masker) and quantization noise (maskee). When coding stereo or multi-channel audio signals, the perceptual model needs to consider this

Encoder:

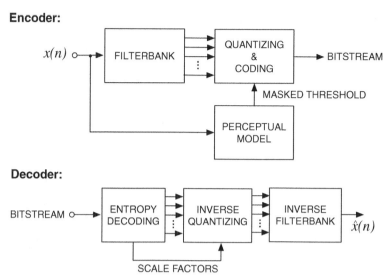

Figure 2.8 Generic perceptual audio encoder and decoder. Encoder: the input signal is decomposed into a number of sub-bands. The sub-band signals are quantized and coded. A perceptual model controls the quantization error such that it is just below the masked threshold and thus not perceptible. Decoder: the inverse entropy coder recovers subband quantization indices and quantization parameters ('scale factors'). After inverse quantization the inverse filterbank is applied to recover the audio signal.

interaural dependence of the masked threshold. This dependence is often described by means of the *binaural masking level difference* (BMLD) [26].

Today many proprietary perceptual audio coders, such as AC-3 (Dolby) [88], ATRAC (Sony) [258], PAC (Bell Laboratories, Lucent Technologies) [243] are available and in some cases (AC-3) widely used. International standards-based perceptual audio coders are widely used, such as MPEG-1 Layer 3 (MP3) [40, 136, 249] and MPEG-2 AAC (used for the Apple iTunes music store) [32, 40, 104, 136, 137, 249]. When requiring that the coded audio signal can not be distinguished from the original audio signal, these perceptual audio coders are able to reduce the bitrate of CD audio signals by a factor of about 10. When higher compression ratios are required, the audio bandwidth needs to be reduced or coding distortions will exceed the masked threshold.

2.3.4 Parametric audio coding

Perceptual audio coders (either sub-band or transform based) with coding mechanisms as described above perform only sub-optimally when high compression ratios need to be achieved. This is in most cases caused by the fact that the amount of bits required to keep the quantization noise below the masked threshold is insufficient, resulting in audible noise, so-called spectral holes, or a reduction in signal bandwidth. *Parametric*

Encoder:

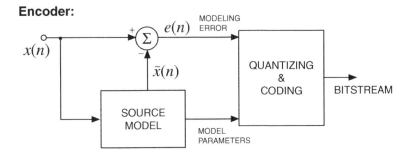

Decoder:

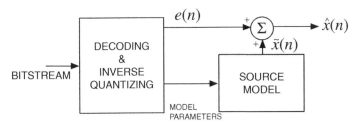

Figure 2.9   A generic parametric encoder and decoder. Encoder: parameters of a source model are estimated adaptively in time for modeling the input signal. The model parameters and often also the modeling error are transmitted to the decoder. Decoder: given the decoded modeling parameters, the source model estimates the original signal. If the error is also coded/transmitted then the source model output is 'corrected' given the modeling error.

audio coders, on the other hand, are usually based on a *source model*, i.e. they make specific assumptions about the signals to be coded and can thus achieve higher compression ratios. Speech coders are a classical example for parametric coding and can achieve very high compression ratios. However, they perform sub-optimally for signals which can not be modeled effectively with the assumed source model. In the following, a number of parametric coding techniques, applicable to audio and speech coding, are described.

A parametric encoder and decoder are illustrated in Figure 2.9. The encoder and decoder operate as follows.

Encoder: a source model predicts the input signal $x(n)$ as a function of modeling parameters. The modeling parameters, and optionally the modeling error $e(n)$, are quantized, coded, and transmitted to the decoder as bitstream. Nonparametric coding techniques may be applied for coding the modeling error, see e.g. [57].

Decoder: the bitstream is parsed and the modeling parameters (and, if used, the modeling error) are decoded and the inverse quantizer is applied. Given the modeling parameters the audio signal is estimated. If the modeling error was also transmitted then the model estimate is improved by adding the modeling error.

Linear predictive coding (LPC) is a standard technique often used in speech coders [4, 163]. In LPC, the source model is composed of an excitation and an all-pole filter. It has been shown that such an all-pole filter is a good model for the human vocal tract thus is suitable for speech coding. LPC has also been applied to wide-band audio signals, see e.g. [31, 110, 111, 242].

Another commonly used parametric coding technique is based on *sinusoidal modeling*, where the source model consists of a number of sinusoids. The parameters to be estimated here are amplitude, frequency, and phase of the sinusoids. Sinusoidal modeling has been applied to speech coding [118, 190] and audio coding [244].

For audio coding, a parametric coder has been proposed which decomposes an audio signal into sinusoids, harmonic components, and noise [76], denoted *harmonic and individual lines plus noise* (HILN) coder [140, 215]. Each of these three 'objects' is represented with a suitable set of parameters. A similar approach, but decomposing an audio signal into sinusoids, noise, and transients, was proposed by den Brinker *et al.* [67].

Parametric coders may not only incorporate a source model, but also a receiver model. Properties of the human auditory system are often considered for determining the choice or coding precision of the modeling parameters. Frequency selectivity properties of the human auditory system were explored for LPC-based audio coding by Härmä *et al.* [111]. A masking model is incorporated into the object-based parametric coder of Edler *et al.* [76] and only signal components which are not masked by the remaining parts of the signal are parameterized and coded. A later version of this coder additionally incorporates a loudness model [216] for giving precedence to signal components of highest loudness.

Most parametric audio and speech coders are only applicable to single-channel audio signals. Recently, the parametric coder proposed by den Brinker *et al.* [67] was extended for coding of stereo signals [233, 234] with parametric spatial audio coding techniques as described in this book.

2.3.5 Combining perceptual and parametric audio coding

Usually, parametric audio coding is applied for very low bitrates, e.g. 4–32 kb/s for mono or stereo audio signals, and nonparametric audio coding for higher bitrates up to transparent coding, e.g. 24–256 kb/s for mono or stereo audio signals. The perceptual audio coder as described above is based on the principle of 'hiding' quantization noise below the masked threshold. It is by design aimed at transparent coding. For medium bitrates, when the amount of quantization noise has to exceed the masked threshold for achieving the desired bitrate, this coding principle may not be optimal. A number of techniques have been proposed for extending perceptual audio coders with parametric coding techniques for improved quality at such medium bitrates.

One such parametric technique applicable for stereo or multi-channel audio coding is *intensity stereo coding* (ISC) [119]. ISC is a parametric technique for coding of audio channel pairs. ISC is illustrated in Figure 2.10. For lower frequencies, the spectra of a pair of audio channels are both coded. At higher frequencies, e.g. >1–3 kHz, only a single spectrum is coded and the spectral differences are represented as intensity differences. The intensity difference parameters contain significantly less information than a spectrum and thus the bitrate is lowered by ISC.

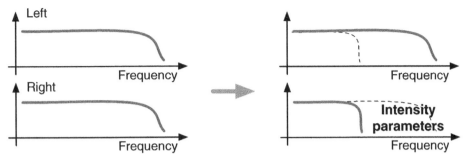

Figure 2.10 Intensity stereo coding (ISC) is applied to a pair of short-time spectra. At higher frequencies (usually above 1–3 kHz) the spectrum pair is represented as a single spectrum plus parameters representing the intensity difference between the channel pair.

Figure 2.11 Spectral bandwidth replication (SBR) represents the higher-frequency spectrum with a few parameters which allow regeneration of a meaningful spectrum at the decoder.

Another parametric technique that has been applied within the framework of perceptual audio coders is denoted *spectral bandwidth replication* (SBR) [68]. SBR is illustrated in Figure 2.11. It applies perceptual audio coding at frequencies below a certain frequency (e.g. <4–12 kHz) and parametric coding above that frequency. Very little information is used to parameterize the upper spectrum and the decoder uses information from the lower spectrum to generate a perceptually meaningful upper spectrum. Ideas related to SBR were already introduced by Makhoul and Berouti [186] in the context of speech coding.

2.4 Matrix surround

Matrix systems enabled wide spread use of surround sound. First in movie theaters as Dolby Stereo and later in homes as Dolby Surround [129]. One reason for the success of matrix systems is its stereo backward compatibility. Matrix systems 'encode' multiple audio channels into two stereo compatible audio channels. Given a matrix encoded stereo signal, a matrix decoder is applied to approximate the original multi-channel audio signal.

Early matrix systems were used to encode four audio channels into a stereo signal, denoted 4-2-4 systems [75]. Modern matrix systems, such as Dolby Prologic II [69], are capable of encoding 5.1 surround audio signals, if desired, including the LFE channel. More recently, also matrix systems for 6.1 and 7.1 signals have been introduced (e.g. Dolby Prologic IIx).

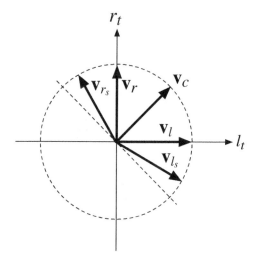

Figure 2.12   The vectors for encoding the five main channels of a 5.1 signal to two channels.

In the following, a simple 5-2-5 matrix system is described for encoding the various channels of a 5.1 surround audio signal. The audio channels are encoded as

$$
\begin{bmatrix} l_t(t) \\ r_t(t) \end{bmatrix} = \begin{bmatrix} 1 & 0 & \frac{1}{\sqrt{2}} & -j\sqrt{\frac{2}{3}} & -j\sqrt{\frac{1}{3}} \\ 0 & 1 & \frac{1}{\sqrt{2}} & j\sqrt{\frac{1}{3}} & j\sqrt{\frac{2}{3}} \end{bmatrix} \begin{bmatrix} l_f(t) \\ r_f(t) \\ c(t) \\ l_s(t) \\ r_s(t) \end{bmatrix}
\tag{2.3}
$$

where $l_f(t)$, $r_f(t)$, $c(t)$, $l_s(t)$, $r_s(t)$ denote the front left, front right, front center, rear left, and rear right audio channels, respectively. The j denotes a 90° phase shift.

Alternatively, (2.3) can be written as

$$
\begin{bmatrix} l_t(t) \\ r_t(t) \end{bmatrix} = \begin{bmatrix} 1 \\ 0 \end{bmatrix} l_f(t) + \begin{bmatrix} 0 \\ 1 \end{bmatrix} r_f(t) + \begin{bmatrix} \frac{1}{\sqrt{2}} \\ \frac{1}{\sqrt{2}} \end{bmatrix} c(t) + j \begin{bmatrix} -\sqrt{\frac{2}{3}} \\ \sqrt{\frac{1}{3}} \end{bmatrix} l_s(t)
$$

$$
+ j \begin{bmatrix} -\sqrt{\frac{1}{3}} \\ \sqrt{\frac{2}{3}} \end{bmatrix} r_s(t)
$$

$$
= \mathbf{v}_l l_f(t) + \mathbf{v}_r r_f(t) + \mathbf{v}_c c(t) + \mathbf{v}_{l_s} l_s(t) + \mathbf{v}_{r_s} r_s(t) ,
\tag{2.4}
$$

where the five vectors are unit vectors mapping the five input channels to the two output channels. The vectors are shown relative to $l_t(t)$ and $r_t(t)$ in Figure 2.12.

The simplest way of decoding a matrix encoded signal is to apply a 'passive decoder matrix'. Formulated in terms of the previously defined vectors, the matrix decoded channels are:

$$
\hat{l}_f(t) = [l_t(t) \ r_t(t)] \mathbf{v}_l
$$

$$\hat{r}_f(t) = [l_t(t) \ r_t(t)] \, \mathbf{v}_r$$

$$\hat{c}(t) = [l_t(t) \ r_t(t)] \, \mathbf{v}_c$$

$$\hat{l}_s(t) = [l_t(t) \ r_t(t)] \, \mathbf{v}_{l_s}$$

$$\hat{r}_s(t) = [l_t(t) \ r_t(t)] \, \mathbf{v}_{r_s} \tag{2.5}$$

Each output channel is equal to the projection of the matrix encoded signal onto the corresponding matrix encoding vector.

As can easily be verified, the described encoding/decoding strategy perfectly reconstructs a channel in a scenario where only a single channel is active. However, the other channels are not zero in this case, i.e. there is a lot of crosstalk. The situation when only one channel is active can easily be detected. In such a situation one could set all the other audio channels to zero and thus improve the channel separation. This is the principle idea behind 'active matrix decoders' [106].

Often the rear channels of matrix systems are low-pass filtered and delayed. The low-pass filtering mimics late reverberation which has high frequencies attenuated due to more absorption of sound in air at high frequencies. The rear channels are delayed to prevent that signal components are 'jumping' between front and back when the matrix decoder can not well separate the channels. Due to the precedence effect (Section 3.4.1) front/back correlated signal components will always be perceived from the front if the back channels are delayed. It is especially important for dialogue in movie sound tracks to firmly always stay in front.

While very practical, since matrix surround enables multi-channel audio, based on a stereo down-mix, matrix surround provides only very limited surround sound quality. For most surround sound material with multiple concurrently active sources and ambience, matrix surround impairs the audio quality in terms of a loss of ambience (due to loss of channel independence) and limited localization (due to high channel crosstalk). Spatial audio coding achieves much higher audio quality by means of 'guided' up-mixing of the down-mix to the original number of channels. The quality of Spatial Audio Coding is more like discrete surround than matrix surround.

2.5 Conclusions

In this Chapter background knowledge has been reviewed. Signals processed by spatial audio coding and processing are usually played back over stereo or multi-channel surround loudspeaker systems, headphones, or headphones by means of binaural audio playback. The specific playback system is implicitly or explicitly considered by spatial audio coding and processing. Further, conventional audio signal representation and coding has been reviewed since spatial audio coding aims at enabling lower bitrate representation for stereo or multi-channel signals than the conventional techniques. Matrix surround has been described due to its similarity with spatial audio coding in terms of representing a surround audio signal using a down-mix signal. The next chapter describes background information, related to spatial hearing, which spatial audio coding and processing are based on.

3

Spatial Hearing

3.1 Introduction

Similarly to the way humans perceive a visual image, humans are also able to perceive an *auditory spatial image*. The different objects which are part of the auditory spatial image are denoted *auditory objects*. For example, if a listener listens to a musical performance, the auditory objects are the different instruments which are playing. In most listening situations, the perceived directions of auditory objects correspond well to the directions of the physical *sound sources* emitting the sounds that are associated with the corresponding auditory objects. This is a necessity in order that the perceived auditory spatial image corresponds to the physical surroundings of a listener.

Figure 3.1 illustrates the perception of the auditory spatial image for a listener being in a performance with three sound sources in a room (left). For each source an auditory object is perceived (1 in Figure 3.1) with a specific position and width. The frontal auditory spatial image has also a total extent (2 in Figure 3.1) which is not necessarily directly associated with the auditory objects. The impression of being within the sound field is denoted listener envelopment (3 in Figure 3.1). The different attributes of the auditory spatial image, auditory object location/width and listener envelopment, have been discussed by Rumsey [228] in the context of spatial quality evaluation.

The chapter is organized as follows. A description of the physiology of the human auditory system is followed by an introduction of spatial hearing phenomena. The important case of spatial hearing in rooms is treated in detail. Limitations of the auditory system are described and a model is presented which attempts to explain the remarkable properties of the human auditory system in terms of perceiving complex auditory spatial images given a mix of source signals and reflections at the ear entrances. The chapter is concluded with a discussion on how the described spatial hearing knowledge relates to and enables spatial audio processing and coding.

3.2 Physiology of the human hearing system

This section briefly describes the most important parts of the human auditory system. For more detailed information the reader is referred to dedicated textbooks, e.g. [226].

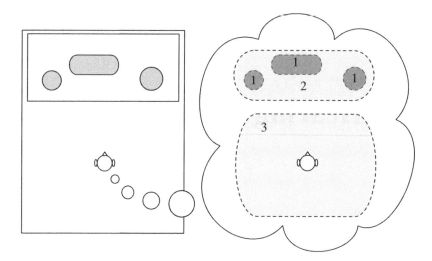

Figure 3.1 Illustration of the perception of the auditory spatial image. Perception of auditory objects (1); extent of the auditory spatial image (2); and listener envelopment (3).

Outer and middle ear

Changes in the acoustical pressure are received by the outer ear, which consists of a visible part (the pinna) and a canal leading to the eardrum (the tympanic membrane). Attached to the eardrum are the middle ear bones (ossicles): the malleus (the outermost), the incus and the stapes. The foot of the stapes is connected to the oval window, which is part of the inner ear. The major function of the middle ear is to ensure the efficient transfer of sound from the air to the fluids of the cochlea. It acts as an impedance-matching device that improves sound transmission.

Inner ear

The cochlea within the inner ear is composed of a bony labyrinth. It is partitioned into three fluid-filled tubes. These tubes are separated by membranes: between the first (scala vestibuli) and second (scala media) there is Reissner's membrane. The scala media is in turn separated from the lower space (scala tympani) by the basilar membrane.

The so-called organ of Corti spirals on the basilar membrane. Two types of sensory cells can be differentiated by their position within the organ of Corti: inner hair cells (IHCs) and outer hair cells (OHCs). The inner hair cells make up a single row of receptors, while the outer hair cells are much greater in number and are often organized in three rows.

The main task of the cochlea is to analyze sound in terms of its intensity, timing and frequency content and to convert mechanical vibrations to neural spikes. The mechanical movements of the ossicles produce displacement in the cochlear fluids and the basilar membrane. The resulting vibration of the basilar membrane takes the form of waves that travel away from the stapes. For a pure tone, the wave's amplitude reaches a maximum at a certain position along the basilar membrane and then collapses. The position of the maximum amplitude depends on the frequency of the tone; high frequencies have a maximum amplitude toward the stapes and low frequencies toward the apex. Hence the cochlea can be seen as a spectrum analyzer, comprising a set of band-pass filters.

The vibrations of the basilar membrane result in deflection of the sensory stereocilia on top of the hair cells. This deflection opens and closes mechanoelectrical transduction channels, producing a sound-induced change in the current. The change in the voltage across the membrane of the sensory neurons results in the release of neurochemicals and stimulation of neurons in IHCs. The OHCs, on the other hand, the voltage controls a molecular force generation mechanism, also known as the cochlear amplifier. These forces increase the vibration of the basilar membrane.

Auditory nerve

The auditory nerve transmits the spikes generated by the IHCs to the cochlear nucleus. For stimulation with single tones at levels above threshold, the average response rate of the auditory nerve increases roughly linearly with the logarithm of the sound pressure over a limited intensity range. At the onset of a tone, the auditory nerve fibers show initially a very high discharge peak, followed by a lower steady-state discharge rate. The difference between the initial (relatively high) and steady-state rate is referred to as adaptation. This adaptation occurs in three stages: the initial stage has a time constant of a few milliseconds, and is followed by a second stage with a longer time constant in the order of tens of milliseconds. A third, even slower stage lasts tens of seconds.

A very prominent feature for responses to low-frequency tones is the presence of multiple peaks spaced at time intervals corresponding to the period of the stimulus frequency. Thus, the temporal structure of the firing rate reflects the periodicity of the stimulus. This is referred to as phase locking [20, 155, 160, 269]. With increasing frequency, the strength of this phenomenon decreases.

Cochlear nucleus

The auditory nerve enters the cochlear nucleus and divides into two branches: an ascending branch that innervates the anteroventral cochlear nucleus (AVCN) and a descending branch that innervates both the posteroventral (PVCN) and the dorsal nucleus (DCN).

Auditory brainstem

At this and subsequent levels of the auditory system, the information present in the auditory nerve undergoes important transformations. Information from the two ears is integrated to extract binaural cues, and significant monaural features are extracted from the input. The superior olivary complex (SOC) is the first place at which afferents from the two cochlear nuclei converge. In most mammals two major types of binaural neurons are found in the SOC. Both the medial and lateral superior olives (MSO and LSO, respectively) receive binaural input and contain neurons sensitive to interaural time differences (ITDs).

Cells in the MSO receive excitatory input from the cochlear nucleus of both sides and are therefore designated excitatory-excitatory (EE). These cells are sometimes referred to as *concidence detectors* and their behavior can be compared to a cross-correlator. Their discharge rate in response to binaural stimulation depends on the interaural time difference (ITD) and, at favorable ITDs, i.e. when exhibiting maximum response, typically exceeds the sum of the responses for either ear alone. This favorable ITD is referred to as the cell's *best delay*. If a given neuron is activated by different frequencies, the different periodic discharge curves appear to reach a maximum amplitude for the same interaural delay of the stimulus. This delay is referred to as the cell's *characteristic delay* and provides an estimate of the difference in travel time from each ear to the coincidence detector. Conceptually, a set of EE-type neurons with different characteristic delays can be compared to a cross-correlation function [152].

A second subgroup of cells in the lateral superior olive (LSO) and a subgroup of cells in the inferior colliculus (IC) are excited by the signals from one ear and inhibited by the signals from the other ear [7, 8, 33, 158, 175, 189, 204, 224]. The cells in the LSO are typically excited by the ipsilateral ear and inhibited by the contralateral ear and are therefore classified as EI-type (excitation–inhibition) cells. For neurons situated in the IC the excitatory and inhibitory channels are typically reversed and these cells are classified as IE-type cells. The opposite influence of the two ears makes these cells sensitive to interaural level differences (ILDs). With increasing inhibitory level, the neuron's activity decreases up to a certain level where its activity is completely inhibited. The ILD necessary to completely inhibit the cell's response varies across neurons [205, 206, 257]. We refer to the minimum interaural intensity difference needed to completely inhibit the activity as the neuron's *characteristic ILD*. There are some suggestive data for the LSO [206, 257] and for the IC [133] that the ILD sensitivity of EI-type neurons reflects the differences in threshold between the excitatory and inhibitory inputs that innervate each EI-type cell. In addition to ILD sensitivity, EI-type cells have been reported to exhibit ITD sensitivity as well [157, 158, 205].

It is however uncertain to what extent EI-type neurons contribute to binaural hearing phenomena in humans. It is estimated that ITD sensitive IE units comprise only 12% of low-frequency units in the IC [204]. Furthermore, anatomical studies revealed that the LSO in humans is much less well developed than in various other animals [200].

3.3 Spatial hearing basics

In the previous section, it was described how various stages of the auditory system respond to ITDs, ILDs and their resemblance to the cross-correlation function. In this section, the relation between physical attributes of sound sources and these localization cues will be described.

3.3.1 Spatial hearing with one sound source

The simplest listening scenario is when there is one sound source in *free-field*. Free-field denotes an open space with no physical objects from which sound is reflected. *Anechoic chambers* are rooms frequently used for experimentation under free-field-like conditions. Due to their highly sound absorbent walls there are virtually no reflections, similarly to free-field. *Localization* denotes the relation between the location of an auditory object and one or more attributes of a *sound event*. A sound event denotes sound sources and their corresponding signals responsible for the perception of the auditory object. For example, localization may describe the relation between the direction of a sound source and the direction of the corresponding auditory object. *Localization blur* denotes the smallest change in one or more attributes of a sound event such that a change in location of the auditory object is perceived. For sources in the *horizontal plane*, localization blur with respect to direction is smallest for sources in the forward direction of a listener. It is slightly larger for sources behind the listener and largest for sources to the sides. In other words, the precision with which a listener can discriminate the direction of a source in the horizontal plane is best if the source is in the front and worst if a source is on the side.

In order to understand how the auditory system discriminates the direction of a source, the properties of the signals at the ear entrances have to be considered, i.e. the signals available to the auditory system. Most generally, the ear input signals can be viewed as being filtered versions of the source signal. The filters modeling the path of sound from a source to the left and right ear entrances are commonly referred to as *head-related transfer functions* (HRTFs). Figure 3.2(a) illustrates the left and right HRTFs, h_1 and h_2, for a source at angle ϕ. For each source direction different HRTFs need to be used for

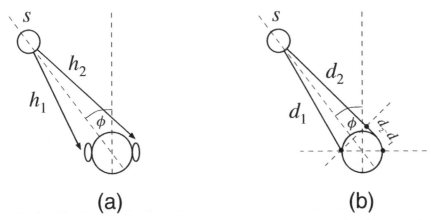

(a)   (b)

Figure 3.2   (a) Paths of a source to the ear entrances modeled with HRTFs; (b) a more intuitive view relates the source angle to a path distance difference resulting in an arrival time difference at the ear entrances. Additionally, head shadowing results in an intensity difference between the ear entrance signals as a function of the source angle.

modeling the ear entrance signals. A more intuitive, but only approximately valid view for the relation between the source angle ϕ and the ear entrance signals is illustrated in Figure 3.2(b). The difference in length of the paths to the ear entrances, $d_2 - d_1$, can be expressed as a function of the source angle ϕ. As a result of the different path lengths, there is a difference in arrival time between both ear entrances. The most simple formula for the difference in path length between the left and right ear is the 'sine law' for spatial hearing as proposed by von Hornbostel and Wertheimer [266],

$$\Delta d = \kappa \sin \phi \qquad \text{with} \qquad \kappa = 21 \text{ cm} \qquad (3.1)$$

where κ is the distance of the two microphones modeling the two ear entrances. Since the effect of the curved path around the head is ignored, κ is chosen larger than the actual distance between the ears. Another limitation of the sine law is that *head shadowing* is ignored, i.e. the effect of the head on the intensities of the ear entrance signals is not considered. Several improved formulas were introduced to account for the curved path of sound around the head. An overview of different path length difference formulas has been given by Blauert [26].

As a result of the path length difference from the source to the two ear entrances, there is a difference in arrival times of sound at the left and right ears, denoted *interaural time difference* (ITD). Note that the physiologically relevant range of ITD is about ±1 ms. This range is related to distance between the ears, by e.g. (3.1), and the resulting physically possible range of ITDs.

Additionally, the shadowing of the head results in an intensity difference of the left and right ear entrance signals, denoted *interaural level difference* (ILD). For example, a source to the left of a listener results in a higher intensity of the signal at the left ear than at the right ear.

Diffraction, reflection, and resonance effects caused by the head, torso, and the external ears of the listener result in that ITD and ILD not only depend on the source angle ϕ, but also on the source signal. Nevertheless, if ITD and ILD are considered as a function of frequency, it is a reasonable approximation to say that the source angle solely determines ITD and ILD as implied by data shown by Gaik [90]. When only considering frontal directions ($-90° \le \phi \le 90°$) the source angle ϕ approximately causally determines ITD and ILD. However, for each frontal direction there is a corresponding direction in the back of the listener resulting in a similar ITD–ILD pair. Moreover, if only ITDs are considered, a cone of positions exist with virtually equal ITDs at all positions of the cone. This concept is referred to as the *cone of confusion*. Thus, the auditory system needs to rely on other cues for resolving this front/back ambiguity. Examples of such cues are head movement cues, visual cues, and spectral cues (different frequencies are emphasized or attenuated depending on the elevation of a sound source) [26].

3.3.2 Ear entrance signal properties and lateralization

The previous discussion implies that ITD and ILD are ear entrance signal properties which provide to the auditory system information about the direction of a sound source in the horizontal plane. A specific ITD–ILD pair can be associated with the source direction (when disregarding the front/back ambiguity). With headphones, the ear entrance signals

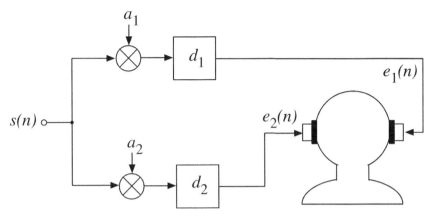

Figure 3.3 Experimental setup for generating coherent left and right ear entrance signals with specific ITD and ILD.

are (ideally) equal to the signals given to the left and right transducers of the headphones. Therefore, it is possible to evaluate the effect of ITD and ILD independently of each other with experiments carried out with headphones. Figure 3.3 shows an experimental setup for generating coherent left and right ear entrance signals, $e_1(n)$ and $e_2(n)$, given a single audio signal $s(n)$. ITD is determined by the delays and equal to $d_2 - d_1$ and ILD is determined by the scale factors a_1 and a_2, and expressed in dB is $20 \log_{10}(a_2/a_1)$.

Figure 3.4(a) illustrates perceived auditory objects for different ITD and ILD [26, 107, 229, 284] for two coherent left and right headphone signals, as generated by the scheme shown in Figure 3.3. When left and right headphone signals are coherent, have the same level (ILD $= 0$), and no delay difference (ITD $= 0$), an auditory object appears in the center between the left and right ears of a listener. More specifically, the auditory object appears in the center of the frontal section of the upper half of the head of a listener, as illustrated by Region 1 in Figure 3.4(a). By increasing the level on one side, e.g. right, the auditory object moves to that side as illustrated by Region 2 in Figure 3.4(a). In the extreme case, when only the signal on the left is active, the auditory object appears at the left side as illustrated by Region 3 in Figure 3.4(a). ITD can be used similarly to control the position of the auditory object. For headphone playback, a subject's task is usually restricted to identifying the lateral displacement of the projection of the auditory object to the straight line connecting the ear entrances. The relationship between the lateral displacement of the auditory object and attributes of the ear entrance signals is denoted *lateralization*.

So far, only the case of coherent left and right ear input signals was considered. Another ear entrance signal property that is considered in this discussion is a measure for the degree of 'similarity' between the left and right ear entrance signals, denoted *interaural coherence* (IC). IC here is defined as the maximum value of the normalized cross-correlation function,

$$\text{IC} = \max_{d} \frac{\sum_{n=-\infty}^{\infty} e_1(n)e_2(n+d)}{\sqrt{\sum_{n=-\infty}^{\infty} e_1^2(n) \sum_{n=-\infty}^{\infty} e_2^2(n+d)}} \qquad (3.2)$$

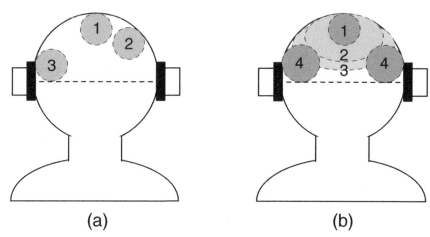

Figure 3.4 (a) ILD and ITD between a pair of headphone signals determine the location of the auditory object which appears in the frontal section of the upper head; (b) the width of the auditory object increases (1–3) as the interaural coherence (IC) between the left and right headphone signals decreases, until two distinct auditory objects appear at the sides (4).

where delays d corresponding to a range of ± 1 ms are considered, which is the physiologically plausible range for ITD, as discussed in Section 3.3.1. IC as defined has a range between zero and one. IC $= 1$ means that two signals are coherent (signals are equal with possibly a different scaling and delay) and IC $= 0$ means that the signals are independent. IC may also be defined as the signed value of the normalized cross-correlation function with the largest magnitude, resulting in a range of values between minus one and one. A value of minus one then means that the signals are identical, but with a different sign (phase inverted).

When two identical signals (IC $= 1$) are emitted by the two transducers of the headphones, a relatively compact auditory object is perceived. For wideband signals the width of the auditory object increases as the IC between the headphone signals decreases until two distinct auditory objects are perceived at the sides, as illustrated in Figure 3.4(b) [58].

Conclusively, one can say that it is possible to control the lateralization of an auditory object by choosing ITD and ILD. Furthermore, the width of the auditory object is related to IC.

3.3.3 Sound source localization

The usual result of lateralization experiments is that the auditory objects are perceived inside the head, somewhere between the left and right ear. One of the reasons for the fact that these stimuli are not externalized is that the single frequency-independent ILD or ITD is a poor representation of the acoustical signals in the free field. The waveforms of sound sources in the real world are filtered by the pinna, head and torso of the listener, resulting in an intricate frequency dependence of the ITD and ILD. For the ILD, this frequency dependence is visualized in Figure 3.5. The ILD fluctuates around zero for an

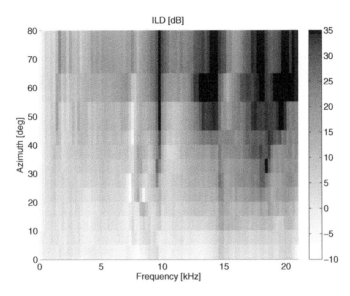

Figure 3.5   Interaural level difference as a function of frequency for various azimuth angles.

azimuth of zero. For increasing azimuth, the ILD tends to increase as well. However, this increase is stronger at high frequencies than at low frequencies. This is caused by the fact that the acoustic shadow of the head is more effective for shorter wavelengths (i.e., higher frequencies).

The observation that ILDs vary stronger with position at high frequencies than at low frequencies, and the fact that human subjects are sensitive to ITDs as well, resulted in the *duplex theory* formulated by Lord Rayleigh in 1907 [251]. This theory states that sound source localization is facilitated by interaural level differences at high frequencies and by interaural time differences at low frequencies.

Besides causing a frequency dependence of the ITD and ILD cues, reflections and diffraction also result in pronounced notches and peaks in the signal spectra themselves [272]. These specific properties of the magnitude spectra facilitate sound source localization in the *vertical* plane. The frequencies at which these features occur depend on the elevation of the sound source. This is visualized in Figure 3.6. The magnitude spectrum in dB is given as a function of the elevation of a sound source in the front of the listener (azimuth = 0°). Especially for frequencies above approximately 8 kHz, differences in the magnitude spectrum of up to 30 dB can occur, depending on the elevation.

Due to differences in the shape of individual heads and ears, the exact spectral features in HRTF spectra differ from individual to individual. When positioning sound sources by HRTF convolution, it is therefore important to apply *individual* HRTFs. If individual HRTFs are used, subjects are not able to discriminate between real and virtual sound sources presented over headphones [115, 177, 273]. If *non-individualized* HRTFs are used, however, subjects report poor elevation accuracy and front–back confusions [271, 275]. Some attempts have been made to increase localization performance with non-individualized HRTFs by emphasizing the pinna effects [288] or the interaural differences [73].

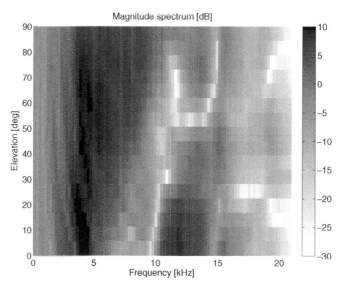

Figure 3.6 HRTF magnitude spectrum as a function of the elevation of a sound source at zero degrees azimuth.

So far, only elevation and azimuth were considered, but there is evidence that spatial cues also depend on the distance of a sound source [53, 240], especially for close-range positions.

3.3.4 Two sound sources: summing localization

Spatial hearing with two sound sources has a high practical relevance because stereo loudspeaker listening depends on perceptual phenomena related to two sound sources (the two sources are the two loudspeakers in this case). Also for multi-channel loudspeaker listening the two-source case is relevant since it is based on similar phenomena.

Previously, the ear entrance signal properties ITD and ILD were related to source angle. Then the perceptual effect of ITD, ILD, and IC cues was discussed. For two sources at a distance (e.g. loudspeaker pair), ITD, ILD, and IC are determined by the HRTFs of both sources and by the specific source signals. Nevertheless, it is interesting to assess the effect of cues similar to ITD, ILD, and IC, but relative to the source signals and not ear entrance signals. To distinguish between these same properties considered either between the two ear entrance signals or two source signals, respectively, the latter are denoted *inter-channel time difference* (ICTD), *inter-channel level difference* (ICLD), and *inter-channel coherence* (ICC). For headphone playback, ITD, ILD, and IC are (ideally) the same as ICTD, ICLD, and ICC. In the following a few phenomena related to ICTD, ICLD, and ICC are reviewed for two sources located in the front of a listener.

Figure 3.7(a) illustrates the location of the perceived auditory objects for different ICLD for two coherent source signals [26]. When left and right source signals are coherent (ICC = 1), have the same level (ICLD = 0), and no delay difference (ICTD = 0), an

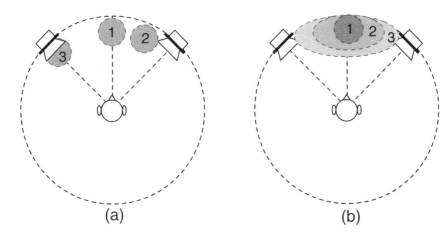

Figure 3.7   (a) ICTD and ICLD between a pair of coherent source signals determine the location of the auditory object which appears between the two sources; (b) the width of the auditory object increases (1–3) as the IC between left and right source signals decreases.

auditory object appears in the center between the two sources, as illustrated by Region 1 in Figure 3.7(a). By increasing the level on one side (hence employing the concept of amplitude panning, see Chapter 2), e.g. right, the auditory object moves to that side, as illustrated by Region 2 in Figure 3.7(a). In the extreme case, when only the signal on the left is active, the auditory object appears at the left source position as is illustrated by Region 3 in Figure 3.7(b). ICTD can be used similarly to control the position of the auditory object (i.e., time panning is employed). This principle of controlling the location of an auditory object between a source pair is also applicable when the source pair is not in front of the listener. However, some restrictions apply for sources to the sides of a listener [211, 256]. There is an upper limit for the angle between such a source pair beyond which localization of auditory objects between the sources degrades.

When coherent wideband signals (ICC = 1) are simultaneously emitted by a pair of sources, a relatively compact auditory object is perceived. When the ICC is reduced between these signals, the width of the auditory object increases [26], as illustrated in Figure 3.7(b). The phenomenon of summing localization, e.g. an auditory object appearing between a pair of frontal loudspeakers, is based on the fact that ITD and ILD cues evoked at the ears crudely approximate the dominating cues that would appear if a physical source were located at the direction of the auditory object. The mutual role of ITDs and ILDs is often characterized with time–intensity trading ratios [26] or in the form of the classic duplex theory [251]: ITD cues dominate localization at low frequencies and ILD cues at high frequencies. The relevance of ITD and ILD cues as a function of frequency is illustrated in Figure 3.8.

The insight that when signals with specific properties are emitted by two sources the direction of the auditory object can be controlled is of high relevance for applications. It is this property, which makes stereo audio playback possible. With two appropriately placed loudspeakers, the illusion of auditory objects at any direction between the two loudspeakers can be generated. (More on this topic is discussed in Section 2.2).

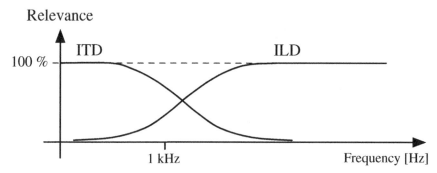

Figure 3.8 The duplex theory states that ITD dominate localization at low frequencies and ILD at high frequencies. The transition between ITD and ILD dominance is about 1–2 kHz.

Another relevance of these phenomena is that for loudspeaker playback and headphone playback similar cues can be used for controlling the location of an auditory object. This is the basis, which makes it possible to generate signal pairs which evoke related illusions in terms of relative auditory object location for both loudspeaker and headphone playback. If this were not the case, there would be a need for different signals, depending on whether a listener uses loudspeakers or headphones.

3.3.5 Superposition of signals each evoking one auditory object

Above, only phenomena were described where a single auditory object appears. However, an auditory spatial image often consists of a number of auditory objects distributed in (perceived) space. Consider the case of two *independent* sound sources. In this context, independent sources denote independent signals, e.g. different speech signals or instru-ments. Over a short period of time (e.g. 20 ms) such signals may be correlated, but when observed over a sufficiently long period of time (e.g. a few seconds) such signals are statistically independent. When one source emits a signal, a corresponding auditory object is perceived at the direction of that source (left scenario in Figure 3.9). When another independent source at a different location emits a signal, another corresponding auditory object is perceived from the direction of that second source (middle scenario in Figure 3.9). When both sources emit their independent signals simultaneously, usually two auditory objects are perceived from the two directions of the two sources, in which the position of each source independently is determined by ICLD and ICTD parameters. From the linearity property of HRTFs it follows that in the latter scenario the ear entrance signals are equal to the sum of the two ear entrance signals of the two one-active-source scenarios (as indicated in the right scenario in Figure 3.9). More generally speaking, the sum of a number of ear entrance signals associated with independent sources usually results in distinct auditory objects for each of the sources. In many cases, the directions of the resulting auditory objects correspond to the directions of the sources.

This perceptual feature is not only a remarkable ability of the auditory system, but also a necessity for formation of auditory spatial images corresponding to the physical

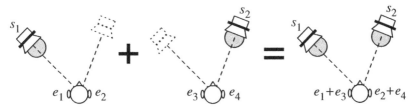

Figure 3.9 When two independent sources are concurrently active (right), usually two auditory objects appear at the same directions where auditory objects appear for each source emitting a signal individually (left and middle).

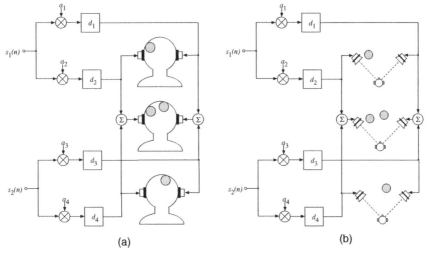

Figure 3.10 By adding two different pairs of signals corresponding to single auditory objects, the composite signal pair usually results in two auditory objects at the same locations as the auditory objects corresponding to each signal pair for (a) headphone playback and (b) loudspeaker playback.

surroundings of a listener. In Section 3.6, a model is described aiming at explaining this phenomenon, i.e. how the auditory system is able to localize sources given the complex mix of signals at the ear entrances in multi-source scenarios.

The principle described for the superposition of ear entrance signals also holds for signal pairs generated as illustrated in Figure 3.3. For example, if a number of such signal pairs are generated for different independent base signals, then the composite signal pair usually results in perceived auditory objects at the locations of the auditory objects appearing when the signal pairs are played back separately. From the linearity property of HRTFs this follows not only for headphone playback, but also for two sources emitting these signals. This property is illustrated for headphones and sources in Figure 3.10(a) and (b), respectively. *Mixing* techniques for generating stereo signals given a multitrack recording (separately recorded instruments) are based on the described principle.

3.4 Spatial hearing in rooms

3.4.1 Source localization in the presence of reflections: the precedence effect

Usually, the direct sound of a source reaches the ears earlier than the reflections of the same sound because the indirect path associated with a reflection is longer than the direct path from the source to the ears. The *precedence effect* describes a number of phenomena related to the auditory system's ability to resolve the direction of a source in the presence of one or more reflections by favoring the 'first wavefront' over successively arriving reflections. That is, the directional perception of reflections arriving within a few milliseconds after the direct sound is suppressed and the direct sound and these reflections are 'fused' into one single auditory object at the direction of the direct sound. Extensive reviews of the precedence effect have been given by Zurek [291], Blauert [26], and Litovsky *et al.* [183].

A typical precedence effect experiment is illustrated in Figure 3.11. The signals given to two loudspeakers are illustrated in Figure 3.11(a). The signal x_1 contains pulses repeating at regular intervals τ_p. The same pulse train is contained in signal x_2, but slightly delayed by τ_e. Typical values for τ_p and τ_e are 400 and 5 ms, respectively. When listening to these signals over a standard stereo setup a listener will perceive only one auditory object at the position of the loudspeaker which emits x_1.

Figure 3.12 illustrates the three phases of the precedence effect. (I) The directional perception of a pair of stimuli with an interstimulus delay shorter than 1 ms is called summing localization (Section 3.3.4). The weight of the lagging stimulus reduces with increasing delay up to approximately 1 ms. (II) For delays greater than that the leading sound dominates the localization judgment. (III) Echo threshold refers to the delay where the fusion breaks apart. Depending on stimulus properties and individual listeners, thresholds of 2–50 ms have been reported in the literature [183]. The previously mentioned

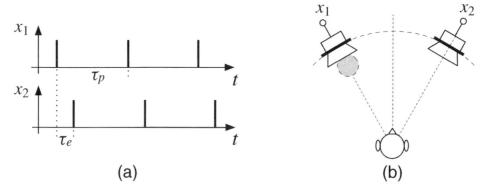

(a)   (b)

Figure 3.11   A typical precedence effect experiment: (a) signals given to the loudspeakers; (b) an auditory object is perceived at the position of the leading-signal loudspeaker.

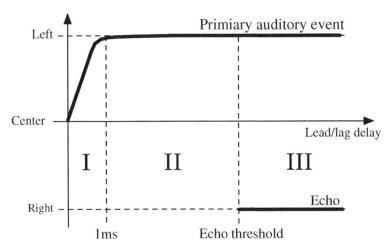

Figure 3.12   The three phases of the precedence effect: (I) summing localization; (II) precedence effect; (III) localization of primary auditory object and echo.

auditory model [87] not only attempts to explain localization of concurrent sources, but also localization of sources in the presence of reflections, i.e. the precedence effect.

Figure 3.13(a) shows the echo threshold for speech signals as investigated by Damaske [65]. The echo threshold for noise pulses of different lengths are shown in Figure 3.13(b) [194]. The considered lengths of the noise pulses are 10, 30, and 100 ms.

3.4.2 Spatial impression

So far the discussion has mostly focused on the attribute of perceived direction or lateralization of auditory objects. One exception was the discussion of the role IC and ICC play for signals in determining the width of the auditory object. In the following, other attributes related to auditory objects and the auditory spatial image are briefly discussed. These attributes mostly depend on the properties of reflections relative to the direct sound.

Coloration

The first early reflections up to about 20 ms later than the direct sound can cause timbral coloration due to a 'comb-filter' effect which attenuates and amplifies frequency components in a frequency-periodic pattern.

An example for the effect of early reflections is illustrated in Figure 3.14. An impulse response h_1 with a direct sound and a single reflection after 7 ms, is shown in the top left panel of Figure 3.14. The corresponding magnitude spectrum is shown in the top right panel, showing that this corresponds to a 'comb-filter', i.e. frequencies at regular intervals are removed from a signal when filtered with h_1. The bottom two panels of Figure 3.14 illustrate another example with several reflections of different strengths.

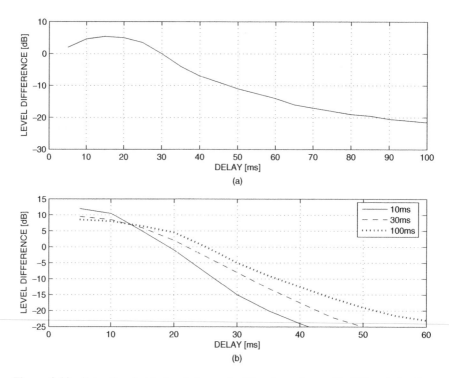

Figure 3.13   Echo thresholds for: (a) speech; (b) noise pulses with different durations.

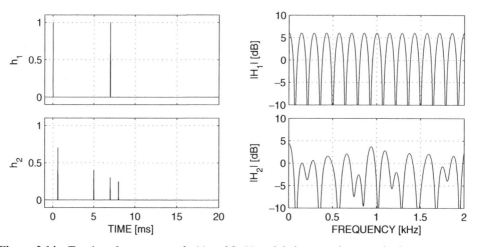

Figure 3.14   Two impulse responses $h_1(t)$ and $h_2(t)$ and their respective magnitude spectra $H_1(\omega)$ and $H_2(\omega)$.

Distance of auditory object

In free field, the following two-ear entrance signal attributes change as a function of source distance: power of signal reaching the ears and high-frequency content (air absorption). Additionally, for sources close to the head a source distance change causes a change in ILD across all frequencies [52]. There is evidence that the overall level of sound reaching the ears provides potent distance information. For a source for which a listener knows its likely level of emitted sound, such as speech, the overall sound level at the ear entrances provides an absolute distance cue [192, 193]. However, in situations when a listener does not expect a source to have a certain emitting level, overall sound level at the ear entrances can not be used for judging absolute distance. In such a situation, overall level provides only a relative cue [62].

On the other hand, in a reverberant environment there is more information available to the auditory system. The reverberation time and the timing of the first reflections contain information about the size of a space and the distance to the surfaces, thus giving an indication about the expected range of source distances. Thus it is not surprising that many researchers have argued that for relatively distant sources the ratio of the power of direct to reflected sound is a reliable distance cue, see e.g. [50, 192, 193].

Distance cues and their importance for generating artificial auditory spatial images have been discussed by Shinn-Cunningham [238]. It is argued that for real-world listening conditions and headphone playback, it is most important to consider level and reverberation cues.

Width of auditory objects

Barron and Marshall [6] found that lateral reflections from $\pm 90°$ cause the greatest spatial impression. The closer the direction of the reflections is to the median plane the less is the resulting spatial impression. The spatial impression caused by a pair of early reflections in the range of reflection delays between about 5 and 80 ms is approximately constant. Based on this, Barron and Marshall proposed a physical measure called *lateral fraction* (LF) for measuring spatial impression. The lateral fraction is the ratio of the lateral sound energy to the total sound energy that arrived within the first 80 ms after the arrival of the direct sound

$$\mathrm{LF}_5^{80} = \frac{\int_{5\ \mathrm{ms}}^{80\ \mathrm{ms}} h^2(t)\cos^2\alpha(t)\,\mathrm{d}t}{\int_{0\ \mathrm{ms}}^{80\ \mathrm{ms}} h^2(t)\,\mathrm{d}t} \tag{3.3}$$

where $h(t)$ is the impulse response and $\alpha(t)$ is the angle of arrival of the reflection at time t relative to the side as illustrated in Figure 3.15. A dipole microphone pointing towards the side can be used to measure $h(t)\cos\alpha(t)$. The response of such a dipole microphone is indicated in Figure 3.15.

The lateral fraction measure is mostly associated with the width of the auditory object. More recent studies found that lateral reflections from $\pm 90°$ are not optimal for creating greatest spatial impression at all frequencies [3, 203, 241], i.e. at certain frequencies reflections arriving from other directions than the side create most spatial impression.

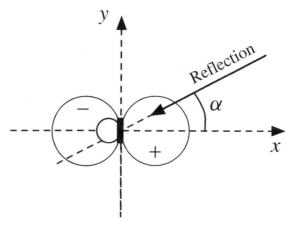

Figure 3.15 Definition of reflection direction for the computation of lateral fraction and late lateral fraction.

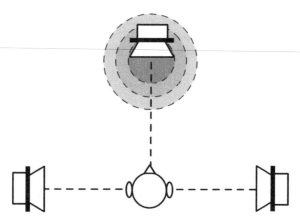

Figure 3.16 Early reflections emitted from the side loudspeakers have the effect of widening the auditory object. The shaded area indicates the perceived auditory object.

An experimental setup for emulating early lateral reflections is illustrated in Figure 3.16. The direct sound is emitted from the center loudspeaker while independent early reflections are emitted from the left and right loudspeakers. The width of the auditory object increases as the relative strength of the early lateral reflections is increased.

Listener envelopment

More than 80 ms after the arrival of the direct sound, reflections tend to contribute more to the perception of the environment than to the auditory object itself. This is manifested in a sense of 'envelopment' or 'spaciousness of the environment', frequently denoted

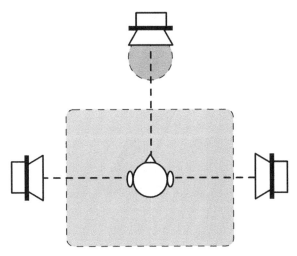

Figure 3.17 Late reflections emitted from the side loudspeakers relate more to the environment than the auditory object itself. This is denoted listener envelopment. The shaded areas indicate the perceived auditory objects.

listener envelopment [201]. Such a situation occurs for example in a concert hall, where late reverberation arrives at the listener's ears from all directions.

Bradley and Soulodre [39] extended the research of Barron and Marshall by adding more early reflections and a (late) 'reverberation tail'. From a number of experiments they concluded that a similar measure as the lateral fraction for early reflections may also be applicable to reverberation. This measure relates more to listener envelopment than width of auditory objects. They termed this measure *late lateral fraction*:

$$\mathrm{LF}_{80}^{\infty} = \frac{\int_{80\ \mathrm{ms}}^{\infty} h^2(t)\cos^2\alpha(t)\,\mathrm{d}t}{\int_{0\ \mathrm{ms}}^{\infty} h^2(t)\,\mathrm{d}t} \qquad (3.4)$$

Late lateral reflections can be emulated with a setup as shown in Figure 3.17. The direct sound is emitted from the center loudspeaker while independent late reflections are emitted from the left and right loudspeakers. The sense of listener envelopment increases as the relative strength of the late lateral reflections is increased, while the width of the auditory object is expected to be hardly affected.

Interaural cross-correlation coefficient

The previously described measures, lateral fraction and late lateral energy fraction, relate properties of rooms (early and late reflections) to the perceptual phenomena of width of auditory objects and listener envelopment. Another class of physical measures relates properties of the signals at the ear entrances to such attributes. In the following a few such measures are reviewed.

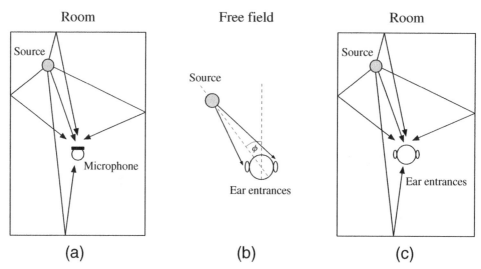

Figure 3.18 (a) Room impulse response (RIR); (b) left and right head-related transfer function (HRTF); (c) left and right binaural room impulse response (BRIR).

In order to represent properties of the ear entrance signals, *binaural room impulse responses* (BRIRs) are considered. Recall that a *room impulse response* (RIR) models the path from a source to an observation point in a room and a *head-related transfer function* (HRTF) models the path from a source to an ear entrance in free field. A BRIR is a linear filter modeling the multiple paths sound travels through in a room before reaching an ear entrance as direct sound and reflections. Usually HRTFs and BRIRs are considered in pairs, one for the left and right ear entrance, respectively. Figure 3.18 illustrates how RIR, a pair of HRTFs, and a pair of BRIRs are defined.

As implied by the results presented in Sections 3.3.2 and 3.3.4, IC and ICC are related to the width of auditory objects. An artificial experience related to listener envelopment can be evoked by emitting independent noise signals with the same level from loudspeakers distributed all around a listener, as illustrated in Figure 3.19. When the ICC between source signal pairs is increased, the width of the auditory object surrounding the listener decreases [64]. IC and ICC are in many cases directly related, i.e. lower ICC between a loudspeaker pair results in lower IC between the ear entrance signals [173]. Thus both, IC and ICC, seem to be related to auditory object width and listener envelopment. Similarly to the case of relating room properties to auditory object width and envelopment, IC can be related to these two properties by computing it relative to the early and late part of BRIRs. These two measures are often denoted early and late *interaural cross-correlation coefficient*, IACC(E) and IACC(L), respectively [38, 203]. The IACC is defined as

$$\text{IACC} = \max_\tau \Phi(\tau) \qquad (3.5)$$

where

$$\Phi(\tau) = \frac{\int_{T_1}^{T_2} h_l(t) h_r(t+\tau)\, dt}{\sqrt{\int_{T_1}^{T_2} h_l^2(t)\, dt \int_{T_1}^{T_2} h_r^2(t)\, dt}} \qquad (3.6)$$

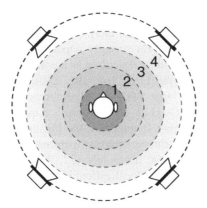

Figure 3.19　For multi-loudspeaker playback the auditory object surrounding the listener increases in width as the ICC between the signals decreases.

Table 3.1　Definition of interaural cross-correlation coefficient measures.

Measure	BRIR segment (ms)
IACC(E)	0–80
IACC(L)	80–1000
IACC(T)	0–1000

and $\tau \in [-1\ 1]$ ms. The early, late, and total IACC use different limits of the integration, T_1 and T_2, in (3.6) as shown in Table 3.1.

Despite the fact that IACC and measures like lateral fraction and lateral energy fraction seem very different, they are often similarly influenced by lateral reflections [174]. A time-based division, by considering early and late reflections (e.g. up to 80 ms and later reflections) for measuring auditory object width and envelopment, is not always suitable since both influence each other to a certain degree.

3.5 Limitations of the human auditory system

3.5.1 Just-noticeable differences in interaural cues

Although the human auditory system is capable of estimating interaural cues with significant accuracy, it has several known limitations as well. For example, interaural cues have to change by a certain amount in order to be detectable. Such minimum required change is often referred to as *just-noticeable difference* (JND) or *threshold*.

For example, the just-noticeable change in ILD amounts to approximately 0.5–1 dB and is roughly constant over frequency and stimulus level [101, 123, 191, 282]. If the reference ILD increases, ILD thresholds increase also. For reference ILDs of 9 dB, the

ILD threshold is about 1.2 dB, and for a reference ILD of 15 dB, the ILD threshold amounts between 1.5 and 2 dB [196, 225, 285].

The sensitivity to changes in ITDs strongly depends on frequency. For frequencies below 1000 Hz, this sensitivity can be described as a constant interaural phase difference (IPD) sensitivity of about 0.05 rad [153, 165, 191, 286]. The reference ITD has some effect on the ITD thresholds: large ITDs in the reference condition tend to decrease sensitivity to changes in the ITDs [123, 283]. There is almost no effect of stimulus level on ITD sensitivity [295]. At higher frequencies, the binaural auditory system is not able to detect time differences in the fine-structure waveforms. However, time differences in the envelopes can be detected quite accurately [22, 259]. Despite this high-frequency sensitivity, ITD-based sound source localization is dominated by low-frequency cues [24, 25].

The sensitivity to changes in the interaural coherence (IC) strongly depends on the reference coherence. For a reference coherence of +1, changes of about 0.002 can be perceived, while for a reference coherence around 0, the change in coherence must be about 100 times larger to be perceptible [63, 89, 178, 222]. The sensitivity to changes in interaural coherence is practically independent of stimulus level, as long as the stimulus is sufficiently above the absolute threshold [109]. At high frequencies, the *envelope* coherence seems to be the relevant descriptor of the spatial diffuseness [19, 20].

The threshold values described above are typical for spatial properties that exist for a prolonged time (i.e., 300–400 ms). If the duration is smaller, thresholds generally increase. For example, if the duration of the (change in) ILD and ITD in a stimulus is decreased from 310 to 17 ms, the thresholds may increase by up to a factor of 4 [21]. Interaural coherence sensitivity also strongly depends on the duration [277, 278, 294]. It is often assumed that the increased sensitivity for longer durations results from temporal integration properties of the auditory system.

There is, however, one important exception in which the auditory system does not seem to integrate spatial information across time. In reverberant rooms, the perceived location of a sound source is dominated by the first 2 milliseconds of the onset of the sound source, while the remaining signal is largely discarded in terms of spatial cues. This phenomenon is referred to as 'the law of the first wavefront' or 'precedence effect' [183, 239, 268, 289], which is discussed in Section 3.4.1.

3.5.2 Spectro-temporal decomposition

Extensive psychophysical research (cf. [125, 168, 263]) and efforts to model the binaural auditory system (cf. [46, 61, 90, 182, 248]) have suggested that the human auditory system extracts spatial cues as a function of time and frequency. To be more specific, there is considerable evidence that the binaural auditory system renders its binaural cues in a set of frequency bands, without having the possibility to acquire these properties at a finer frequency resolution. This spectral resolution of the binaural auditory system can be described by a filter bank with filter bandwidths that follow the ERB (equivalent rectangular bandwidth) scale [98, 108, 166].

The limited temporal resolution at which the auditory system can track binaural localization cues is often referred to as 'binaural sluggishness', and the associated time constants are between 30 and 100 ms [125, 167]. Although the auditory system is not able to *follow* ILDs and ITDs that vary quickly over time, this does not mean that listeners are not able to *detect the presence* of quickly varying cues. Slowly varying ILDs and/or ITDs result in a movement of the perceived sound source location, while fast changes in binaural cues lead to a percept of 'spatial diffuseness', or a reduced 'compactness' [26]. In other words, there exists a transition from distinct localization cues (if ITDs or ILDs remain constant within the temporal analysis window of the auditory system) to a decrease in interaural coherence (if ITDs or ILDs vary considerably within the temporal analysis window). Despite the fact that the perceived 'quality' of the presented stimulus depends on the movement speed of the binaural cues, it has been shown that the *detectability* of ILDs and ITDs is practically *independent* of the variation speed [45]. This sensitivity of human listeners to time-varying changes in binaural cues can also be described by sensitivity to changes in the maximum of the cross-correlation function (e.g., the *coherence*) of the incoming waveforms [20, 60, 152, 247]. In fact, there is considerable evidence that the sensitivity to changes in any of the binaural cues is the basis of the phenomenon of binaural masking level difference (BMLD) [71, 102, 103, 154, 191]. The BMLD reflects a change in the detection threshold of a signal (the maskee) in the presence of another masking signal (the masker), if the spatial properties of the masker are different from those of the signal (maskee). For example, when a masking tone is presented in phase to both ears, and a pure-tone maskee is presented out-of-phase to each ear simultaneously (a so-called $NoS\pi$ condition), the threshold level for detecting the signal is generally lower than for the case when both the masker and the maskee are presented in phase (NoSo condition) [102, 281]. In this condition, the addition of the out-of-phase maskee results in the generation of a *constant* ITD or ILD, depending on the phase angle between masker and maskee, which is used as a cue for detection. For the NoSo condition, this cue is absent and hence the detection thresholds for the signal are in many cases higher (up to 25 dB).

If the masker consists of noise, the interaural cues due to the addition of the out-of-phase signal will randomly fluctuate across time, depending on the bandwidth of the masking noise [45, 290]. The sensitivity to a decrease in coherence due to the signal results in lower detection thresholds than for an in-phase signal [44, 124, 260, 292].

Recently, it has been demonstrated that the concept of 'spatial diffuseness' depends mostly on the interaural coherence value itself and is relatively unaffected by the temporal fine-structure details of the waveforms within the temporal integration time of the binaural auditory system. For example, van de Par *et al.* [261] measured the detectability and discriminability of inter-aurally out-of-phase test signals presented in an inter-aurally in-phase masker. The subjects were perfectly able to *detect* the presence of the out-of-phase test signal, but they had great difficulty in *discriminating* different test signal types (i.e. noise vs harmonic tone complexes). In a second series of experiments van de Par *et al.* [262] a harmonic tone complex and a noise signal were presented simultaneously to two ears. In one ear, the harmonic tone complex was lagging while in the other ear, the noise was lagging. In other words, the harmonic tone complex (when presented in isolation) would have been lateralized to one ear, while the noise would have been lateralized to the other ear. Surprisingly, subjects had great difficulty in detecting a left/right swap of the stimulus as long as no distinct temporal structure was present in the harmonic tone complex. On the other hand, subjects could easily discriminate between the left/right lateralized stimulus and a condition without any time lags (i.e., both the noise and harmonic tone complex presented diotically).

These observations suggest that detection in a BMLD condition is indeed based on a change in binaural cues, and that the underlying signals that cause the change in binaural cues can not be isolated from the masking signal on a time–frequency grid that is more accurate than the time–frequency resolution of the binaural auditory system. In other words, *detection* of a change in binaural cues does not imply *identification* of the signals that cause the change.

3.5.3 Localization accuracy of single sources

The combined effect of spectral and interaural cues results in the ability of human listeners to discriminate between different positions in the horizontal plane with an accuracy of 1–10° [161, 195, 221]. Absolute localization tasks usually result in a lower accuracy between 2 and 30° [53, 187, 221, 273]. In the vertical direction, localization accuracy amounts to about 4 to 20 degrees [187, 208, 221, 273]. It has also been shown that *changes* in the localization cues, as long as the movement of the sound source is relatively slow [209], increase our ability to localize sound sources [208, 275].

3.5.4 Localization accuracy of concurrent sources

Localization accuracy in the presence of concurrent sound from different directions has been investigated by several authors. A detailed review is given by Blauert [26]. The effect of independent distracters on the localization of a target sound has been recently studied by Good and Gilkey [99], Good *et al.* [100], Lorenzi *et al.* [185], Hawley *et al.* [116], Drullman and Bronkhorst [70], Langendijk *et al.* [176], Braasch and Hartung [36], and Braasch [35]. The results of these studies generally imply that the localization of the target is either not affected or only slightly degraded by introducing one or two simultaneous distracters at the same overall level as the target. When the number of distracters is increased or the target-to-distracter ratio (T/D) is reduced, the localization performance begins to degrade. However, for most configurations of a target and a single distracter in the frontal horizontal plane, the accuracy stays very good down to a target level only a few dB above the threshold of detection [99, 100, 185]. An exception to these results is the outcome of the experiment of Braasch [35], where two incoherent noises with exactly the same envelope were most of the time not individually localizable.

3.5.5 Localization accuracy when reflections are present

Localization accuracy within rooms has been studied by Hartmann [112], Rakerd and Hartmann [219, 220], and Hartmann and Rakerd [114] (see also a review by Hartmann [113]). Overall, in these experiments the localization performance was slightly degraded by the presence of reflections. Interestingly, using slow-onset sinusoidal tones and a single reflecting surface, Rakerd and Hartmann [219] found that the precedence effect sometimes failed completely. In a follow-up study, the relative contribution of the direct sound and the steady-state interaural cues to the localization judgment was found to depend on the onset rate of the tones [220]. Nevertheless, absence of an attack transient did not prevent the correct localization of a broadband noise stimulus [112]. Giguère and Abel [96] reported similar findings for noise with the bandwidth reduced to one-third octave. Rise/decay time had little effect on localization performance except for the lowest center frequency (500 Hz), while increasing the reverberation time decreased the localization accuracy. Braasch *et al.* [37] investigated the bandwidth dependence further, finding that the precedence effect started to fail when the bandwidth of noise centered at 500 Hz was reduced to 100 Hz.

3.6 Source localization in complex listening situations

In addition to the spatial hearing knowledge presented so far, for spatial audio processing and coding it is useful to understand how the auditory system determines the locations of sources in complex listening scenarios. The relation between interaural time difference (ITD) and interaural level difference (ILD) and source direction in free-field is obvious and a conclusion that the auditory system discriminates the source direction as a function of ITD and ILD is in this case rather plausible. However, some phenomena have also been described for which it is not obvious how the auditory system processes ear entrance signal properties for localization of sound sources. For example, when a number of sources are concurrently active, ITD and ILD are likely to be time varying and in many cases their values do not correspond directly to source directions. In an enclosed space, when sound from sources not only reaches the ears of a listener directly, but also indirectly from different directions, the matter of localization of the sources becomes even more complicated. Playback of 'real-world' stereo and multi-channel audio signals usually mimics listening to multiple concurrently active sources in rooms. For the task of spatial audio processing or designing a coding scheme for spatial audio it is helpful to understand which signal properties are important to the auditory system for source localization, source width, and envelopment perception. These properties need to be maintained when coding stereo and multi-channel audio signals in order that the auditory spatial image of the coded audio signal corresponds to the auditory spatial image of the original signal.

While single-source localization in the presence of reflections can be at least partially explained by the precedence effect, the attempt of the model, described in the following, is to explain source localization for the general case of multiple concurrently active sources and reflections. The model is denoted *cue selection model* and is described in greater detail in [87].

3.6.1 Cue selection model

In most listening situations, the perceived directions of auditory objects coincide with the directions of the corresponding physical sound sources. In everyday complex listening scenarios, sound from multiple sources, as well as reflections from the surfaces of the physical surroundings, arrive concurrently from different directions at the ears of a listener. The auditory system not only needs to be able to independently localize the concurrently active sources, but it also needs to be able to suppress the effect of the reflections. The cue selection model, described in the following, qualitatively explains both of these features.

The basic approach of the cue selection model is very straightforward: only ITD and ILD cues occurring at time instants when they represent the direction of one of the sources are selected, while other cues are ignored. The interaural coherence (IC) is used as an indicator for these time instants. More specifically, in many cases by selecting ITD and ILD cues coinciding with IC cues larger than a certain threshold, one obtains a subset of ITD and ILD cues similar to the corresponding cues of each source presented separately in free-field. The cue selection is implemented in the framework of a model that considers a physically and physiologically motivated peripheral stage, whereas the remaining parts are analytically motivated. Fairly standard binaural analysis is used to calculate the instantaneous ITD, ILD, and IC cues.

Model overview

The auditory system features a number of physical, physiological, and psychological processing stages for accomplishing the task of source direction discrimination and ultimately the formation of the auditory spatial image. The structure of a generic model for spatial hearing is illustrated in Figure 3.20. There is little doubt about the first stages of the auditory system, i.e. the physical and physiological functioning of the outer, middle, and inner ear are known and understood to a high degree. However, already the stage of the binaural processor is less well known. Different models have used different approaches to explain various aspects of binaural perception. The majority of proposed localization models are

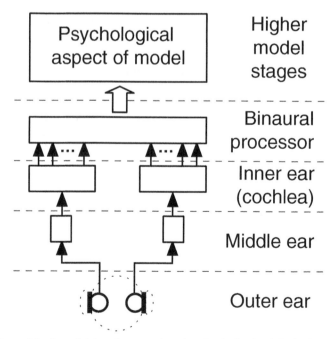

Figure 3.20   A model of spatial hearing covering the physical, physiological, and psychological aspects of the auditory system.

based on analysis of ITD cues using a coincidence structure [152], or a cross-correlation implementation that can be seen as a special case of the coincidence structure. Evidence for cross-correlation-like neural processing has also been found in physiological studies [280]. However, such excitation–excitation (EE) type cells are but one kind of neural units potentially useful for obtaining binaural information (see e.g. the introduction and references of [46] and Section 3.3). With current knowledge, the interaction between the binaural processor and higher level cognitive processes can be addressed only through indirect psychophysical evidence.

Auditory periphery

Transduction of sound from a source to the ears of a listener is modeled by filtering the source signals either with head-related transfer functions (HRTFs) or with measured binaural room impulse responses (BRIRs). HRTF filtering simulates the direction dependent influence of the head and outer ears on the ear input signals. BRIRs additionally include the effect of room reflections in an enclosed space. In multi-source scenarios, each source signal is first filtered with a pair of HRTFs or BRIRs corresponding to the simulated location of the source, and the resulting ear input signals are summed before the next processing stage.

The effect of the middle ear is typically described as a bandpass filter. The frequency analysis of the basilar membrane is simulated by passing the left and right ear signals through a gammatone filterbank [207]. An example of magnitude responses for a set of gammatone filters with center frequencies between about 500 and 2000 Hz are illustrated in Figure 3.21. Each resulting critical band signal is processed using a model of neural transduction, e.g. as described in Bernstein *et al.* [23]. The resulting nerve firing densities at the corresponding left and right ear critical bands are denoted x_1 and x_2.

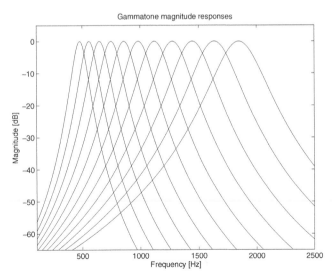

Figure 3.21 Magnitude responses for a set of gammatone filters with center frequencies of 10–20 ERB.

Internal noise is introduced into the model in order to describe the limited accuracy of the auditory system. For this purpose independent Gaussian noise, filtered with the same gammatone filters as the considered critical band signals, is added to each critical band signal before applying the model of neural transduction.

Binaural processor

The cue selection model in principle does not make a specific physiological assumption about the binaural processor. The only assumption is that its output signals (e.g. binaural activity patterns) yield information which can be used by the upper stages of the auditory system for discriminating ITD, ILD, and IC. Given this assumption, the cue selection model computes the ITD, ILD, and IC directly. Note that in this discussion ITD, ILD, and IC are defined with respect to critical band signals after applying the neural transduction. ITD, ILD and IC estimation is part of many existing models for sound source localization and binaural unmasking. Hence the proposed approach can in principle be integrated in many existing models.

The binaural cues are estimated with a running process as a function of time in each critical band. The ITD and IC are estimated from the normalized cross-correlation function $\Phi(n, m)$, where n is the time index and m is the lag index of the normalized cross-correlation function.

Choosing the time constant T for the running process to compute the binaural cues is a difficult task. Studies of binaural detection actually suggest that the auditory system integrates binaural data using a double-sided window with time constants of both sides in the order of 20–60 ms (e.g. [125, 167]). However, a double-sided window with this large time constant will not be able to explain the precedence effect, where the localization of a lead sound should not be influenced by a lagging sound after only a few milliseconds. The difference in the observed behavior for these different cases could be explained by assuming that the auditory system comprises multiple temporal analysis stages that have different time constants, or a system that can adapt its temporal resolution to the task that it has to perform. Here a single-sided exponential time window with a time constant of 10 ms is used, in accordance with the time constant of the temporal inhibition of the model of Lindemann [182].

$\Phi(n, m)$ is evaluated over time lags in the range of $[-1, 1]$ ms. The ITD (in samples) is estimated as the lag of the maximum of the normalized cross-correlation function,

$$\text{ITD}(n) = \arg \max_{m} \Phi(n, m) \qquad (3.7)$$

The IC is defined as the maximum value of the instantaneous normalized cross-correlation function,

$$\text{IC}(n) = \max_{m} \Phi(n, m) \qquad (3.8)$$

This estimate describes the coherence of the left and right ear input signals. In principle, it has a range of $[0, 1]$, where 1 occurs for perfectly coherent left and right critical band signals. The neural transduction introduces a DC offset and thus the values of $\text{IC}(n)$ are typically higher than 0 even for independent (nonzero) critical band signals.

The ILD is computed as the level difference between the left and right critical band signals in dB. Note that, due to the effect of neural transduction, the resulting ILD estimates will be smaller than the level differences between the ear input signals.

Higher model stages

A vast amount of information is available to the upper stages of the auditory system through the signals from the auditory periphery. The focus of the cue selection model lies only in the analysis of the three inter-channel properties between left and right critical band signals that were defined in the previous section: ITD, ILD, and IC. It is assumed that at each time instant n the information about the values of these three signal properties, is available for further processing in the upper stages of the auditory system.

Consider the simple case of a single source in free-field. Whenever there is sufficient signal power, the source direction determines the nearly constant ITD and ILD which appear between each left and right critical band signal. The (average) ITDs and ILDs occurring in this scenario are denoted *free-field cues* in the following. The free-field cues of a source with an azimuthal angle ϕ are denoted ITD_ϕ and ILD_ϕ. It is assumed that this kind of a one-source free-field scenario is the reference for the auditory system. That is, in order for the auditory system to perceive auditory objects at the directions of the sources, it must obtain ITD and/or ILD cues similar to the free-field cues corresponding to each source that is being discriminated. The most straightforward way to achieve this is to select the ITD and ILD cues at time instants when they are similar to the free-field cues. In the following it is shown how this can be done with the help of IC.

When several independent sources are concurrently active in free-field, the resulting cue triplets $\{\text{ILD}(n), \text{ITD}(n), \text{IC}(n)\}$ can be classified into two groups. (1) Cues arising at time instants when only one of the sources has power in that critical band. These cues are similar to the free-field cues (direction is represented in $\{\text{ILD}(n), \text{ITD}(n)\}$, and $\text{IC}(n) \approx 1$). (2) Cues arising when multiple sources have non-negligible power in a critical band. In such a case, the pair $\{\text{ILD}(n), \text{ITD}(n)\}$ does not represent the direction of any single source, unless the superposition of the source signals at the ears of the listener incidentally produces similar cues. Furthermore, when the two sources are assumed to be independent, the cues are fluctuating and $\text{IC}(n) < 1$. These considerations motivate the following method for selecting ITD and ILD cues. Given the set of all cue pairs, $\{\text{ILD}(n), \text{ITD}(n)\}$, only the subset of pairs is considered which occurs simultaneously with an IC larger than a certain threshold, $\text{IC}(n) > \text{IC}_0$. This subset is denoted

$$\{\text{ILD}(n), \text{ITD}(n) | \text{IC}(n) > \text{IC}_0\}. \tag{3.9}$$

The same cue selection method is applicable for deriving the direction of a source while suppressing the directions of one or more reflections. When the 'first wavefront' arrives at the ears of a listener, the evoked ITD and ILD cues are similar to the free-field cues of the source, and $\text{IC}(n) \approx 1$. As soon as the first reflection from a different direction arrives, the superposition of the source signal and the reflection results in cues that do not resemble the free-field cues of either the source or the reflection. At the same time IC reduces to IC <1, since the direct sound and the reflection superimpose as two signal pairs with different ITD and ILD. Thus, IC can be used as an indicator for whether ITD and ILD cues are similar to free-field cues of sources or not (while ignoring cues related to reflections).

For a given IC_0 there are several factors determining how frequently $IC(n) > IC_0$. In addition to the number, strengths, and directions of the sound sources and room reflections, $IC(n)$ depends on the specific source signals and on the critical band being analyzed. In many cases, the larger IC_0 the more similar the selected cues are to the free-field cues. However, there is a strong motivation to choose IC_0 as small as possible while still getting accurate enough ITD and/or ILD cues, because this will lead to the cues being selected more often, and consequently to a larger proportion of the ear input signals contributing to the localization.

It is assumed that the auditory system adapts IC_0 for each specific listening situation, i.e., for each scenario with a constant number of active sources at specific locations in a constant acoustical environment. Since the listening situations do not usually change very quickly, it is assumed that IC_0 is adapted relatively slowly in time. In [87] it is also argued that such an adaptive process may be related to the buildup of the precedence effect. The simulations reported in the following consider only one specific listening situation at a time. Therefore, for each simulation a single constant IC_0 is used.

It is assumed that for each specific listening situation the auditory system adapts IC_0 until ITD and ILD cues representing source directions are obtained by the cue selection.

3.6.2 Simulation examples

As mentioned earlier, the cue selection model assumes that, in order to perceive an auditory object at a certain direction, the auditory system needs to obtain cues similar to the free-field cues corresponding to a source at that direction. In the following, the cue selection model is applied to several stimuli that have been used in previously published psychophysical studies. Both, the selected cues as well as all cues prior to the selection, are illustrated and the implied directions are discussed in relation to the literature. Many more simulations are presented in [87].

Usually, the larger the cue selection threshold c_0, the smaller is the difference between the selected cues and the free-field cues. The choice of c_0 is a compromise between the similarity of the selected cues to the free-field cues and the proportion of the ear input signals contributing to the resulting localization.

Here, application of the cue selection is only considered independently at single critical bands. Except for different values of c_0, the typical behavior appears to be fairly similar at critical bands with different center frequencies. The listening situations are simulated using HRTFs. All simulated sound sources are located in the frontal horizontal plane and the stimuli are aligned to 60 dB SPL averaged over the whole stimulus length.

Independent sources in free-field

A speech source can still be rather accurately localized in the presence of one or more competing other speech sources, Hawley *et al.* [116] and Drullman and Bronkhorst [70]. Thus, to be correct, the cue selection has to yield ITD and ILD cues similar to the free-field cues of each of the speech sources in order to correctly predict the directions of

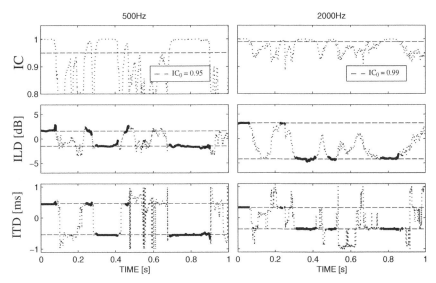

Figure 3.22   IC, ILD, and ITD as a function of time for two independent speech sources at ±40° azimuth. Left column 500 Hz; right column 2 kHz critical band. The cue selection thresholds (top row) and the free-field cues of the sources (middle and bottom rows) are indicated with dashed lines. Selected cues are marked with bold solid lines.

the perceived auditory objects. A simulation was carried out with two concurrent speech sources. The signal of each source consisted of a different phonetically balanced sentence from the Harvard IEEE list [132] recorded by the same male speaker. The two speech sources were simulated at azimuthal angles of ±40°. Figure 3.22 shows the IC, ILD, and ITD as a function of time for the critical bands with center frequencies of 500 Hz and 2 kHz. The free-field cues which would occur with a separate simulation of the sources at the same angles are indicated with the dashed lines. The selected ITD and ILD cues (3.9) are marked with bold solid lines. Thresholds of $c_0 = 0.95$ and $c_0 = 0.99$ were used for the 500 Hz and 2 kHz critical bands, respectively, resulting in 65 and 54% selected signal power. The selected cues are always close to the free-field cues, implying perception of two auditory objects located at the directions of the sources, as reported in the literature. As expected, due to the neural transduction IC has a smaller range at the 2 kHz critical band than at the 500 critical band. Consequently, a larger c_0 is required.

Precedence effect

In the following, the cue selection is illustrated within the context of the precedence effect. In a classical precedence effect experiment, a lead/lag pair of clicks is presented to the listener [26, 183]. The leading click is first emitted from one direction, followed by another identical click from another direction after an *interclick interval* (ICI) of a few milliseconds. For ICIs within a range of about 1 to 10 ms a listener perceives sound only from the direction of the leading click.

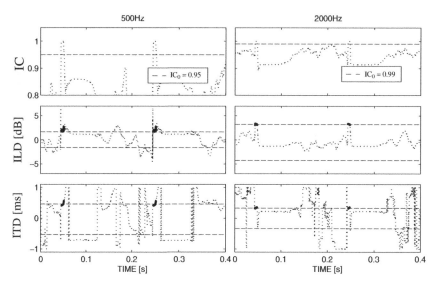

Figure 3.23   IC, ILD, and ITD as a function of time for a lead/lag click-train with a rate of 5 Hz and an ICI of 5 ms. Left column 500 Hz; right column 2 kHz critical band. The cue selection thresholds (top row) and the free-field cues of the sources (middle and bottom rows) are indicated with dashed lines. Selected cues are marked with bold solid lines.

Figure 3.23 shows IC, ILD, and ITD as a function of time for a click train with a rate of 5 Hz analyzed at the critical bands centered at 500 Hz and 2 kHz. The lead source is simulated at $40°$ and the lag at $-40°$ azimuth with an ICI of 5 ms. As expected based on earlier discussion, IC is close to one whenever only the lead sound is within the analysis time window. As soon as the lag reaches the ears of the listener, the superposition of the two clicks reduces the IC. The cues obtained by the selection with $c_0 = 0.95$ for the 500 Hz and $c_0 = 0.985$ for the 2 kHz critical band are shown in the figure, and the free-field cues of both sources are indicated with dashed lines. The selected cues are close to the free-field cues of the leading source and the cues related to the lag are ignored (not selected), as is expected to happen based on psychophysical studies [183]. The fluctuation in the cues before each new click pair is due to the internal noise of the model.

3.7 Conclusions

Spatial audio coding and processing aims at achieving certain goals by exploiting properties of spatial hearing. Spatial hearing and limitations of the human auditory system were discussed. Spatial audio coding and processing manipulate the signal inherent properties which are relevant for spatial hearing. For example, Spatial audio coding achieves data reduction by not representing each signal waveform directly, but by means of a down-mix signal and restoring during decoding only the signal properties which are important for the perception of the auditory spatial image.

4

Spatial Audio Coding

4.1 Introduction

The concept of spatial audio coding is to represent two or more audio channels by means of a down-mix, accompanied by parameters to model the spatial attributes of the original audio signals that are lost by the down-mix process. These 'spatial parameters' capture the perceptually-relevant spatial attributes of an auditory scene and provide means to store, process and reconstruct the original spatial image.

In this chapter the concept of spatial audio coding is explained. The first implementations of spatial audio coding techniques employed a single audio channel as down-mix. This approach is also denoted *binaural cue coding* (BCC). The spatial audio coding approach and concepts using a single audio down-mix channel (BCC) are explained in detail in the current chapter. The extension to multiple down-mix channels is explained in the context of MPEG Surround in Chapter 6.

Figure 4.1 shows a BCC encoder and decoder. As indicated in the figure, the input audio channels $x_c(n)$ ($1 \leq c \leq C$) are down-mixed to one single audio channel $s(n)$, denoted *down-mix signal*. As 'perceptually relevant differences' between the audio channels, inter-channel time difference (ICTD), inter-channel level difference (ICLD), and inter-channel coherence (ICC), are estimated as a function of frequency and time and transmitted as *side information* to the decoder. The decoder generates its output channels $\hat{x}_c(n)$ ($1 \leq c \leq C$) such that ICTD, ICLD, and ICC between the channels approximate those of the original audio signal.

The scheme is able to represent multi-channel audio signals at a bitrate only slightly higher than what is required to represent a mono audio signal. This is so, because the estimated ICTD, ICLD, and ICC between a channel pair contain about two orders of magnitude less information than an audio waveform.

Not only the low bitrate, but also the backward compatibility aspect is of interest. The transmitted down-mix signal corresponds to a mono down-mix of the stereo or multi-channel signal. For receivers that do not support stereo or multi-channel sound

Spatial Audio Processing: MPEG Surround and Other Applications   Jeroen Breebaart and Christof Faller
© 2007 John Wiley & Sons, Ltd

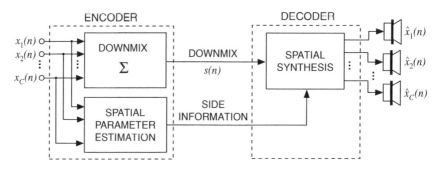

Figure 4.1   Generic scheme for binaural cue coding (BCC).

reproduction, listening to the transmitted down-mix signal is thus a valid method of presenting the audio material on low-profile mono reproduction setups. BCC can therefore also be used to enhance existing services involving the delivery of mono audio material towards multi-channel audio. For example, existing mono audio radio broadcasting systems can be enhanced for stereo or multi-channel playback if the BCC side information can be embedded into the existing transmission channel.

Section 4.2 reviews previously proposed related techniques. BCC is motivated and explained in detail in Section 4.3. This includes a discussion of how ICTD, ICLD, and ICC relate to properties of auditory objects and the auditory spatial image. Multi-channel surround systems often support one or more discrete audio channels for low-frequency effects, denoted LFE channel (for more details see Section 2.2.3). Section 4.4 describes how to apply BCC for efficient coding of LFE channels.

4.2 Related techniques

4.2.1 Pseudostereophonic processes

BCC relies on a synthesis technique which can generate stereo and multi-channel signals given a mono signal. There is a long history of techniques attempting to 'enhance' mono signals to create a spatial impression, i.e. to generate a signal pair or more channels evoking some kind of spatial impression. Such techniques are often called 'pseudostereophonic' processes. Janovsky [151] proposed a scheme where a lowpass filtered version of the mono signal is given to one loudspeaker and a highpass filtered version to the other loudspeaker. Another technique uses complementary comb filters for generating left and right signals [179]. Schroeder [232] proposed the use of allpass filters instead of comb filters resulting in a stereo effect with less coloration artifacts. The use of a reverberation chamber with one loudspeaker emitting the mono signal and two microphones generating left and right signals was described by Schroeder [231] and Lochner and Keet [184]. Another scheme gives the mono signal to both loudspeakers and adds an attenuated and delayed version of the mono signal to one loudspeaker and the same phase-inverted attenuated and delayed signal to the other loudspeaker [179, 180]. Enkl [80] proposed the use of time-variant controllable filters controlled by properties of the mono signal. A more thorough review on these pseudosterephonic processes is given in [26].

In all these techniques, the spatial distribution of the auditory objects is independent of where the sound was originally picked up. The fundamental difference between these early techniques for 'enhancing' mono signals and the technique applied in BCC is that not an arbitrary auditory spatial image is to be created, but an auditory spatial image similar to the auditory spatial image of the original audio signal. For this purpose information about the auditory spatial image must be available. The before-mentioned 'perceptually relevant differences' between the audio channels represent this information.

4.2.2 Intensity stereo coding

Intensity stereo coding (ISC) is a joint-channel coding technique that is part of the ISO/IEC MPEG family of standards [32, 40, 136, 137, 249]. ISC is applied to reduce 'perceptually irrelevant information' of audio channel pairs and originated from [267]. In each coder sub-band that uses ISC, the sub-band signals are replaced by a down-mix signal and a direction angle (azimuth). The azimuth controls the intensity stereo position of the auditory object created at the decoder. Only one azimuth is transmitted for a scale factor band (1024 coder bands are divided into roughly 50 scale factor bands that are spaced proportionally to auditory critical bands). ISC is capable of significantly reducing the bitrate for stereo and multi-channel audio where it is used for channel pairs. However, its application is limited since intolerable distortions can occur if ISC is used for the full bandwidth or for audio signals with a highly dynamic and wide spatial image [119]. Potential improvements of ISC are constrained since the time–frequency resolution is given by the core audio coder and cannot be modified without adding considerable complexity.

Some of the limitations of ISC are overcome by BCC by using different filterbanks for coding of the audio waveform and for parametric coding of spatial cues [13]. Most audio coders use a *modified discrete cosine transform* (MDCT) [188] for coding of audio waveforms. The advantages of using a different filterbank for intensity stereo are reduced aliasing [13] and more flexibility, such as the ability to efficiently synthesize not only intensities (ICLD), but also time delays (ICTD) and coherence (ICC) between the audio channels. Another notable conceptual difference between ISC and BCC is that the latter operates on the fullband audio signal and transmits only a mono time domain signal, whereas ISC does not transmit a true mono fullband signal, but only operates on a certain (high) frequency range.

4.3 Binaural Cue Coding (BCC)

4.3.1 Time–frequency processing

BCC processes audio signals with a certain time and frequency resolution. The frequency resolution used is largely motivated by the frequency resolution of the auditory system (see Chapter 3). Psychoacoustics suggest that spatial perception is most likely based on a critical band representation of the acoustic input signal [26]. This frequency resolution is considered by using an invertible filterbank with sub-bands with bandwidths equal or proportional to the critical bandwidth of the auditory system [98, 293]. The specific time and frequency resolution used for BCC is discussed later in Section 4.3.3.

4.3.2 Down-mixing to one channel

It is important that the transmitted down-mix signal contains all signal components of the input audio signal. The goal is that each signal component is fully maintained. Simple summation of the audio input channels often results in amplification or attenuation of signal components. In other words, the power of signal components in the 'simple' sum is often larger or smaller than the sum of the power of the corresponding signal component of each channel. Therefore, a down-mixing technique is used which *equalizes* the down-mix signal such that the power of signal components in the down-mix signal is approximately the same as the corresponding power in all input channels.

Figure 4.2 shows the down-mixing scheme. The input audio channels $x_c(n)$ ($1 \le c \le C$) are decomposed into a number of sub-bands. One such sub-band is denoted $\tilde{x}_c(k)$ (note that for notational simplicity no sub-band index is used). Since similar processing is independently applied to all sub-bands it is sufficient to describe the processing carried out for one single sub-band. A different time index k is used since usually the sub-band signals are downsampled.

The signals of each sub-band of each input channel are added and then multiplied by a factor $e(k)$

$$\tilde{s}(k) = e(k) \sum_{i=1}^{C} \tilde{x}_i(k) \tag{4.1}$$

The factor $e(k)$ is computed such that

$$\sum_{i=1}^{C} p_{\tilde{x}_i}(k) = e^2(k) p_{\tilde{x}}(k) \tag{4.2}$$

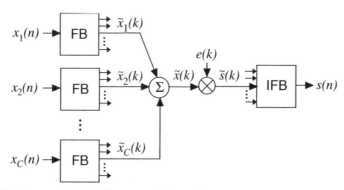

Figure 4.2   The down-mix signal is generated by adding the input channels in a sub-band domain and multiplying the down-mix with a factor in order to preserve signal power. FB denotes filterbank and IFB inverse filterbank. The processing shown is applied independently to each sub-band.

where $p_{\tilde{x}_i}(k)$ is a short-time estimate of the power of $\tilde{x}_i(k)$ at time index k and $p_{\tilde{x}}(k)$ is a short-time estimate of the power of $\sum_{i=1}^{C} \tilde{x}_i(k)$. From (4.2) it follows that

$$e(k) = \sqrt{\frac{\sum_{i=1}^{C} p_{\tilde{x}_i}(k)}{p_{\tilde{x}}(k)}} \qquad (4.3)$$

The equalized sub-bands are transformed back to the time domain, resulting in the down-mix signal $s(n)$ that is transmitted to the BCC decoder.

An example for the effect of the described down-mixing with equalization is illustrated in Figure 4.3. The top two rows show two signals and their respective magnitude spectra. The bottom row shows the sum of the two signals and the magnitude spectra of the 'simple' sum signal and the equalized sum signal as generated with the scheme shown in Figure 4.2. In the range from 500 Hz to about 1 kHz the equalization for this example prevents that the sum signal is significantly attenuated. The top panel in Figure 4.4 shows the same data as shown in the bottom right panel in Figure 4.3. The bottom panel of Figure 4.4 shows the normalized magnitude spectra for down-mix (thin) and equalized down-mix (bold). At each frequency the normalization is relative to the total (sum of) input signal channel power. As desired, with equalization, the normalized magnitude spectrum is 0 dB.

These examples are somewhat artificial since the x_1 and x_2 signals look more like impulse responses (direct sound plus one reflection) than real-world signals. Equalization can be understood as modifying the interactions of impulse responses to prevent attenuation or amplification of signal components due to certain phase interactions.

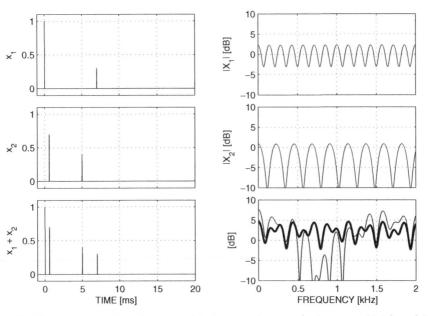

Figure 4.3 Two signals $x_1(n)$ and $x_2(n)$ and their respective magnitude spectra $X_1(f)$ and $X_2(f)$ (top two rows). Down-mix signal $x_1(n) + x_2(n)$ magnitude spectrum (bottom right, thin) and equalized down-mix signal magnitude spectrum (bottom right, bold) are shown.

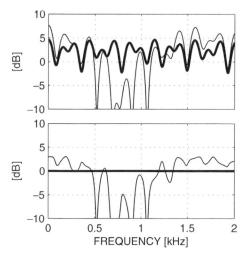

Figure 4.4   Top: down-mix signal $x_1(n) + x_2(n)$ magnitude spectrum (thin) and equalized down-mix signal magnitude spectrum (bold). Bottom: normalized down-mix signal magnitude spectrum (thin) and normalized equalized down-mix signal magnitude spectrum (bold).

4.3.3 'Perceptually relevant differences' between audio channels

The basic operation principle of BCC is to reconstruct a stereo or multi-channel audio signal from the mono down-mix signal that sounds virtually identical to the original stereo or multi-channel content, by reinstating perceptually relevant spatial properties. In Chapter 3 it is explained that in the horizontal plane, ICLD, ICTD and ICC parameters are the most relevant attributes that determine the perceived spatial image. The ICLD and ICTD parameters are associated with the perceived position of a sound source, while the ICC corresponds to a perceived 'width' of one sound source (for example introduced by reflections in a room), and it can indicate unreliable ICLD and ICTD parameters if concurrent sources from different directions are present in the same time/frequency region. In the latter case, the ICLD and ICTD parameters will be time and frequency dependent and follow from interactions of the parameters from each sound source individually (a detailed analysis of ICLD, ICTD and ICC parameters for multiple simultaneous sound sources is given in Chapter 8). But besides reflections or concurrent sound sources, low ICC values may also result from certain effects applied by audio engineers.

The human auditory system has access to ICLD, ICTD and ICC parameters, but its accuracy to analyze these cues is limited. The first limitation exists in the temporal domain. As described in Chapter 3, fast variations of binaural cues can not be tracked by the human auditory system. This behavior seems best described by analysis of a certain 'window' with a typical length of 30–60 ms from which the ICLD, ICTD and ICC parameters are estimated. An important consequence of such an integration window is that very fast variations in either ICLD or ICTD (i.e., variations within one analysis window) result in a decrease of the ICC that is estimated from that window. In other words, fast variations in ICTD and ICLD can be detected, but are perceived as a change in the width rather than fast movements and consequently cannot be tracked in terms of position. Contrary to such an window-averaging model, in precedence-effect situations, the perceived spatial attributes are dominated by the first few milliseconds of a signal onset, ignoring spatial attributes of a significant portion of the remainder of the signal.

A second limitation that was explained in Chapter 3 is the limited frequency resolution to analyze binaural cues. Specifically, on top of the temporal resolution limitations described above, binaural cues seem to be rendered in critical bands only. Moreover, given the finite slope of the so-called auditory filters, significant correlation between binaural cues from adjacent filters is expected, which may loosen the spectral design criteria for BCC schemes somewhat.

A third limitation that can be exploited in BCC is the fact that binaural cues require a certain minimum change in order to be perceived. This property forms the basis for a limited repertoire of binaural cue values, which are closely matched to just-noticeable differences in each cue.

In the current description, filterbanks with sub-bands of bandwidths equal to two times the *equivalent rectangular bandwidth* (ERB) [98] are used. Informal listening revealed that the audio quality of BCC did not notably improve when choosing higher frequency resolution. A lower frequency resolution is favorable since it results in less ICTD, ICLD, and ICC values that need to be transmitted to the decoder and thus in a lower bitrate.

Regarding time-resolution, ICTD, ICLD, and ICC are considered at regular time intervals of about 4–16 ms. Note that using such intervals, the precedence effect is not directly considered. Assuming a classical lead–lag pair of sound stimuli, when the lead and lag fall into a time interval where only one set of cues is synthesized, localization dominance of the lead is not accounted for. Despite this, BCC achieves audio quality reflected in an average MUSHRA score [148] of about 87 ('excellent' audio quality) on average and up to nearly 100 for certain audio signals [12]. Hence when using a fixed update rate of about 4–16 ms, a good trade-off between bit rate and quality is obtained. A dynamic, signal-dependent segmentation process can further improve the performance (see Chapter 5) by lowering the *average* update rate while at the same time providing a high temporal resolution at signal onsets to account for the precedence effect.

The often achieved perceptually small difference between reference signal and synthesized signal implies that cues related to a wide range of auditory spatial image attributes are implicitly considered by synthesizing ICTD, ICLD, and ICC at regular time intervals. In the following, some arguments are given on how ICTD, ICLD, and ICC may relate to a range of auditory spatial image attributes.

Source localization (auditory object direction)

The model for source localization described in Section 3.6 speculates about a possibly important role IC may play for source localization. This includes localization of sources in the presence of concurrent sound and reflections. The validity of this model would in many cases justify the use of ICTD, ICLD, and ICC only at regular time intervals without explicitly considering the precedence effect for real-world audio signals.

Attributes related to reflections

Early reflections up to about 20 ms result in coloration of sources' signals. This coloration effect is different for each audio channel determined by the timing of the early reflections contained in the channel. BCC does not attempt to retrieve the corresponding early reflected sound for each audio channel (which is a source separation problem). However, frequency dependent ICLD synthesis imposes on each output channel the spectral envelope of the original audio signal and thus is able to mimic coloration effects caused by early reflections.

Most perceptual phenomena related to spatial impression seem to be related directly to the nature of reflections that occur following the direct sound. This includes the nature of early reflections up to 80 ms and late reflections beyond 80 ms. Thus it is crucial that the effect of these reflections is mimicked by the synthesized signal.

ICTD and ICLD synthesis ideally result in that each channel of the synthesized output signal has the same temporal and spectral envelope as the original signal. This includes the decay of reverberation (the sum of all reflections is preserved in the transmitted sum signal and ICLD synthesis imposes the desired decay for each audio channel individually). ICC synthesis de-correlates signal components that were originally de-correlated by lateral reflections. Also, there is no need to consider reverberation time explicitly. Blindly synthesizing ICC at each time instant to approximate ICC of the original signal has the desired effect of mimicking different reverberation times, since ICLD synthesis imposes the desired rate of decay.

The most important cues for auditory object distance are overall sound level and direct sound to total reflected sound ratio [238]. Since BCC generates level information and reverberation such that it approaches that of the original signal, auditory object distance cues are represented by considering ICTD, ICLD, and ICC cues.

4.3.4 Estimation of spatial cues

In the following, it is described how ICTD, ICLD, and ICC are estimated. (Depending on the transform that is used to enable sub-band processing, it is sometimes more convenient to analyze inter-channel phase differences (ICPDs) instead of ICTDs, which will be outlined in Chapter 5.) The bitrate required for transmission of these spatial cues is just a few kb/s and thus with BCC it is possible to transmit stereo and multi-channel audio signals at bitrates close to what is required for a single audio channel.

Estimation of ICTD, ICLD, and ICC for stereo signals

The scheme for estimation of ICTD, ICLD, and ICC is shown in Figure 4.5. The following measures are used for ICTD, ICLD, and ICC for corresponding sub-band signals $\tilde{x}_1(k)$ and $\tilde{x}_2(k)$ of two audio channels:

ICTD [samples]:

$$\tau_{12}(k) = \arg\max_d \{\Phi_{12}(d, k)\} \qquad (4.4)$$

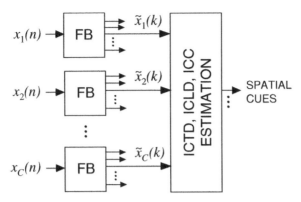

Figure 4.5 The spatial cues, ICTD, ICLD, and ICC are estimated in a sub-band domain. The spatial cue estimation is applied independently to each sub-band.

with a short-time estimate of the normalized cross-correlation function

$$\Phi_{12}(d, k) = \frac{p_{\tilde{x}_1 \tilde{x}_2}(d, k)}{\sqrt{p_{\tilde{x}_1}(k - d_1) p_{\tilde{x}_2}(k - d_2)}} \tag{4.5}$$

where

$$d_1 = \max\{-d, 0\}$$
$$d_2 = \max\{d, 0\} \tag{4.6}$$

and $p_{\tilde{x}_1 \tilde{x}_2}(d, k)$ is a short-time estimate of the mean of $\tilde{x}_1(k - d_1)\tilde{x}_2(k - d_2)$.

ICLD [dB]:

$$\Delta L_{12}(k) = 10 \log_{10} \left(\frac{p_{\tilde{x}_2}(k)}{p_{\tilde{x}_1}(k)} \right) \tag{4.7}$$

ICC:

$$c_{12}(k) = \max_d |\Phi_{12}(d, k)| \tag{4.8}$$

Note that the absolute value of the normalized cross-correlation is considered and $c_{12}(k)$ has a range of $[0, 1]$. Out-of-phase signal pairs can not be represented by these cues as defined. Real-world audio signals only contain phase-inverted signal components in unusual cases and are not explicitly considered here. An alternative ICC/ICTD representation that does allow for out-of-phase signal pairs is outlined in Chapter 6.

Estimation of ICTD, ICLD, and ICC for multi-channel audio signals

The estimation of inter-channel cues between more than 2 channels is somewhat more complex than for the two-channel case as explained above. In principle, $C(C - 1)$ pairs exist to compute parameters from. Due to symmetries in the parameter definition, this will result in $C(C - 1)/2$ unique parameter values (see Figure 4.7(a) for a five-channel

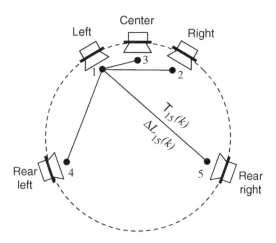

Figure 4.6 ICTD and ICLD are defined between the reference channel 1 and each of the other $C - 1$ channels.

example). For the ICLD, all channel levels are uniquely defined by $C - 1$ ICLD parameters given inter-relations of ICLD parameters and the fact that the overall power should be preserved with respect to the down-mix. For example, in a three-channel case, the ICLD between channel 2 and 3 is fully determined by the ICLDs between channel 1 and 2 on the one hand, and channel 1 and 3 on the other hand. For ICTD and ICC, however, such relations between pair-wise parameters do not exist. However, it was nevertheless found that for the ICTD, $C - 1$ parameters seem sufficient to obtain a perceptually correct spatial image. The ICTD (and ICLD) parameters are defined against one single reference channel. This is illustrated in Figure 4.6 for the case of $C = 5$ channels. $\tau_{1c}(k)$ and $\Delta L_{1c}(k)$ denote the ICTD and ICLD between the reference channel 1 and channel c.

For ICC, a different approach is used. As opposed to using the ICC between all possible channel pairs, it has shown to be sufficient to consider just a few or even a single ICC parameter to indicate the overall coherence or 'diffuseness' of the audio channels. One possibility is to estimate and transmit only ICC cues between the two channels with most energy in each sub-band at each time index. This is illustrated in Figure 4.7(b), when for time instants $k - 1$ and k the channel pairs (3, 4) and (1, 2) are strongest, respectively. The decoder uses this ICC value to determine its de-correlation processing. A different method of reducing the number of ICC parameters is used in MPEG Surround by employing a tree structure of pair-wise channel comparisons as outlined in Chapter 6.

4.3.5 Synthesis of spatial cues

Figure 4.8 shows the scheme which is used in the BCC decoder to generate a stereo or multi-channel audio signal, given the transmitted sum signal plus the spatial cues. The sum signal $s(n)$ is decomposed into sub-bands, where $\tilde{s}(k)$ denotes one such sub-band. For generating the corresponding sub-bands of each of the output channels, delays d_c, scale

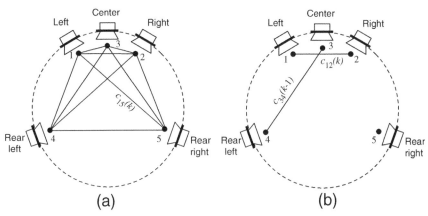

Figure 4.7 Computation of ICC for multi-channel audio signals. (a) In the most general case, ICCs are considered for each sub-band between each possible channel pair; (b) BCC considers for each sub-band at each time instant k, the ICC between the channel pair with the most power in the sub-band considered. In the example shown the channel pair is (3, 4) at time instant $k - 1$ and (1, 2) at time instant k.

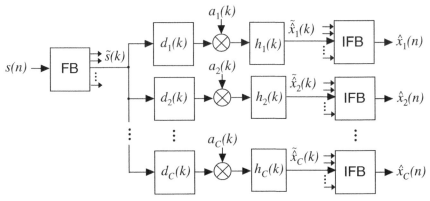

Figure 4.8 ICTD are synthesized by imposing delays, ICLD by scaling, and ICC by applying de-correlation filters. The processing shown is applied independently to each sub-band.

factors a_c, and filters h_c are applied to the corresponding sub-band of the sum signal. (For simplicity of notation, the time index k is ignored in the delays, scale factors, and filters).

ICTD synthesis

The delays are determined by the ICTDs

$$d_c = \begin{cases} -\frac{1}{2}(\max_{2\leq l\leq C} \tau_{1l}(k) + \min_{2\leq l\leq C} \tau_{1l}(k)), & c = 1 \\ \tau_{1c}(k) + d_1, & 2 \leq c \leq C \end{cases} \tag{4.9}$$

The delay for the reference channel, d_1, is computed such that the maximum magnitude of the delays d_c is minimized. The less the sub-band signals are modified, the less danger there is for artifacts to occur. If the sub-band sampling rate does not provide high enough time-resolution for ICTD synthesis, delays can be imposed more precisely by using suitable allpass filters.

ICLD synthesis

In order that the output sub-band signals have desired ICLDs (Equation 4.7) between channel c and the reference channel 1, $\Delta L_{1c}(k)$, the gain factors a_c must satisfy

$$\frac{a_c}{a_1} = 10^{\frac{\Delta L_{1c}(k)}{20}}. \tag{4.10}$$

Additionally, the output sub-bands are normalized such that the sum of the power of all output channels is equal to the power of the input sum signal. Since the total original signal power in each sub-band is preserved in the sum signal (Section 4.3.2), this normalization results in that the absolute sub-band power for each output channel approximates the corresponding power of the original encoder input audio signal. Given these constraints, the scale factors are

$$a_c = \begin{cases} 1/\sqrt{1 + \sum_{i=2}^{C} 10^{\Delta L_{1i}/10}}, & \text{for } c = 1 \\ 10^{\Delta L_{1c}/20} a_1, & \text{otherwise} \end{cases} \tag{4.11}$$

ICC synthesis

The aim is to reduce correlation between the sub-bands after delays and scaling have been applied, without affecting ICTD and ICLD. Generally speaking, this can be achieved by applying filters, h_c in Figure 4.8, in a similar spirit as to generate a certain amount of late reverberation. One point of view for designing the filters h_c in Figure 4.8 is to vary ICTD and ICLD as a function of frequency such that the average value is zero in each sub-band (i.e., the mean cue within one critical band remains unchanged). Figure 4.9

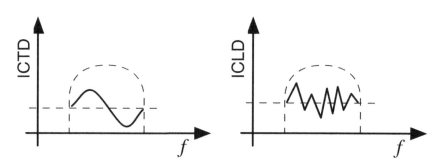

Figure 4.9   ICC is synthesized in sub-bands by varying ICTD and ICLD as a function of frequency.

illustrates how ICTD and ICLD are varied within a sub-band as a function of frequency. The amplitude of ICTD and ICLD variation determines the degree of de-correlation and is controlled as a function of ICC. Note that ICTD are varied smoothly while ICLD are varied randomly. One could vary ICLD as smoothly as ICTD, but this would result in more coloration of the resulting audio signals. Detailed processing for ICTD and ICLD variation as a function of ICC is described in [83].

Another method for synthesizing ICC, particularly suitable for multi-channel ICC synthesis, is described in [82, 218, 233–235]. As a function of time and frequency, specific amounts of artificial (late) reverberation is added to each of the output channels for achieving a desired ICC. Additionally, spectral modification is applied such that the spectral envelope of the resulting signal approaches the spectral envelope of the original audio signal.

4.4 Coding of Low-frequency Effects (LFE) Audio Channels

Commonly used multi-channel surround formats, such as 5.1 surround (Section 2.2.3), use LFE channels. An LFE channel, as defined for the 5.1 standard [150], contains only frequencies up to 120 Hz. In the following the incorporation of the LFE channel in BCC is described. The same principles are applicable to other surround formats.

At frequencies below 120 Hz, six-channel BCC is applied, i.e. all six channels including the LFE channel are coded. At frequencies above 120 Hz, five-channel BCC is applied, i.e. all channels except the LFE channel. The LFE channel is not considered at higher frequencies since it does not contain any signal energy there.

This is implemented specifically by using a filterbank with a lowest sub-band covering 0–120 Hz. For this lowest sub-band the LFE channel is considered and for all other sub-bands the LFE channel is ignored. BCC only considering ICLD is applied to this lowest frequency sub-band since at such low frequencies spatial hearing is limited and the purpose of applying BCC is merely to provide each loudspeaker with the same power as was present in the original audio signal.

Since LFE channels are considered only in the lowest frequency BCC sub-band, at each time instant only a single ICLD parameter per LFE channels is used. Thus the amount of side information does not notably increase by including one or two LFE channels.

4.5 Subjective performance

Although the subjective quality of a BCC scheme may depend on several design choices in the analysis and synthesis stages, as well as the choice of the mono coder, the fact that a parametric model for spatial properties is employed results in two generic, typical observations.

- The subjective quality at low bitrates (typically 12–48 kbps for stereo audio content) for BCC-enabled coders is typically significantly higher than conventional stereo or multi-channel coders.

- With increasing parameter bitrate, the subjective quality saturates at a high, but not fully transparent level due to the limitations of the underlying parametric model.

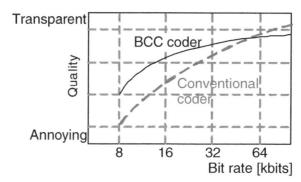

Figure 4.10 Typical performance of a BCC-enhanced coder (solid line) in comparison with a conventional stereo coder (dashed line).

The typical performance characteristic for stereo content and a comparison with a conventional stereo coder is shown in Fig. 4.10. At bitrates below a certain point (in this example about 48 kbit/s) the BCC-enhanced coder outperforms the conventional stereo coder in terms of quality. For higher bitrates, however, the quality for the BCC-enhanced coder tends to saturate. The saturation results from the fact that BCC only restores *statistical* properties of the signals, and does not guarantee a full decoder-side waveform reconstruction. This quality limitation can be resolved by employing the concept of *residual coding* (see Chapter 6).

4.6 Generalization to spatial audio coding

While BCC aims at compressing multi-channel audio by using a single down-mix channel, a different trade-off between compression ratio and audio quality can be achieved by using more than one down-mix channel. A more general scheme, encoding a multi-channel audio signal with an arbitrary number of down-mix channels and spatial cues, has been often denoted spatial audio coding.

In the next chapters, some commercially available audio coders that employ spatial audio coding techniques will be discussed. 'parametric stereo', described in the next chapter, which is an integral part of the aacPlus v2 codec, employs a mono down-mix accompanied by parameters to encode two-channel audio. MPEG Surround, described in Chapter 6, is a very flexible coder in terms of number of input and down-mix channels. An important use case for MPEG Surround is an operation mode based on a stereo down-mix. The use of two down-mix channels results in an improvement in the overall quality and provides backward compatibility with stereo systems, which is an important property for many applications.

5

Parametric Stereo

5.1 Introduction

Parametric stereo (PS) is the first employment of spatial audio coding technology in international standards and commercially available audio codecs. Within MPEG-4, PS is supported in aacPlus v2 (also known as the high-efficiency AAC (HE-AAC) profile), a codec based on (mono) AAC, spectral band replication (SBR) and PS. This codec is currently regarded as the most efficient (stereo) audio coder available today, delivering 'good' quality at bitrates as low as 24–32 kbps, and 'excellent' quality around 48 kbps.

5.1.1 Development and standardization

AacPlus v2 was developed roughly between 2001 and 2004, shortly after the finalization of MPEG-4 Audio version 1 and 2 in 1999–2000. At that point in time, one of the most efficient audio codecs was MPEG-4 AAC, a codec based on transform-domain (MDCT) coefficient quantization, enhanced with perceptual noise substitution (PNS) and long-term prediction (LPT) tools. The performance (or perceptual quality) of MPEG-4 AAC for stereo signals is excellent for bitrates of 96 kbps and higher, but drops rapidly for lower bitrates.

In 2001, MPEG started work to develop new technology to enable further bitrate reduction. Two areas of improved audio coding technology were identified;

- Improved compression efficiency of audio signals or speech signals by means of bandwidth extension, that is forward and backward compatible with existing MPEG-4 technology;

- Improved compression efficiency of high-quality audio signals by means of parametric coding.

Spatial Audio Processing: MPEG Surround and Other Applications   Jeroen Breebaart and Christof Faller
© 2007 John Wiley & Sons, Ltd

Three new tools resulted from these work items:

- SSC (sinusoidal coding) [66, 233, 234], a parametric audio coder based on decomposition of an audio signal into three objects: sinusoids, transients and noise, which can all be described very efficiently using parametric techniques;

- SBR (spectral band replication) [78, 105, 145], a method to efficiently describe the upper bandwidth of an audio signal using a parametric approach;

- PS (parametric stereo) [47], a variant of spatial audio coding that was actually initially developed in the context of SSC [135, 235].

SSC (in combination with PS) reached the final stage of MPEG-4 amendment 2 [135, 143] in mid-2003. However, since PS and SBR could be combined in a very cost-effective manner [214, 217, 235] resulting in an additional boost in coding efficiency for SBR-enabled codecs (i.e., aacPlus [77, 279]), the specific combination of MPEG-4 AAC, SBR and PS was standardized as enhanced aacPlus or aacPlus v2 [217]. Other (application-oriented) standardization bodies such as 3GPP [1] and DVB-H subsequently adopted this technology as well.

5.1.2 AacPlus v2

The general structure of an aacPlus v2 encoder is shown in the top panel of Figure 5.1. The stereo input signal is first processed by a parametric stereo encoder, resulting in a mono audio output signal and (stereo) spatial parameters. The mono signal is subsequently processed by the aacPlus v1 encoder. Finally, a multiplexer combines the spatial parameters and the mono bitstream into a single output bitstream. The spatial parameters are stored in the so-called *ancillary data* part of the core coder bitstream. This part of the bitstream is ignored by legacy (aacPlus v1) codecs, to ensure forward compatibility with existing decoders. In fact, the aacPlus v2 bitstreams are also compatible with MPEG-4 AAC, since both the SBR as well as the PS parameters are stored in the ancillary data part of the AAC bitstream. In other words, any AAC, aacPlus v1 or aacPlus v2 codec can in principle play back any bitstream generated by any of these encoders.

The structure of the aacPlus v2 decoder is shown in the lower panel of Figure 5.1. The decoder basically performs the reverse process of the encoder. The incoming bitstream is first split in a core (aacPlus v1) coder bitstream and a (stereo) spatial parameter bitstream. The core-coder bitstream is subsequently decoded by the aacPlus v1 decoder. Finally, the parametric stereo decoder generates stereo output signals based on the mono input and the transmitted parameters.

In this chapter, the parametric stereo encoder and decoder that are included in aac-Plus v2 will be described in more detail. In international standards such as MPEG, encoders are in many cases not fully specified. In MPEG terminology, it is said that the encoder is *informative*. A simple encoder embodiment is provided as an example, while building a high-quality encoder is up to the developers themselves. On the other hand, the bitstream format and the decoder are fully specified (*normative*). Therefore, the

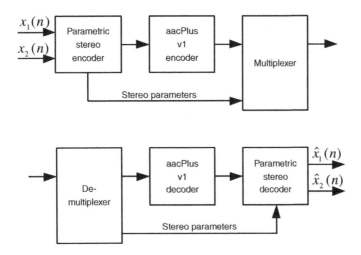

Figure 5.1   Structure of the aacPlus v2 encoder (top panel) and decoder (lower panel).

encoder will be described mostly general terms and concepts, leaving a lot of freedom for the implementer or developer for optimization, while the decoder will be described in more detail. In particular, the specific combination of aacPlus (v1) as core coder and parametric stereo has some interesting complexity advantages when combined in a single system. This will be outlined in more detail in Section 5.5.6.

5.2 Interaction between core coder and spatial audio coding

So far, the core coder and the parametric stereo coder have mostly been described as separate processes. It is well known that core coders result in audible artifacts if the bitrate is relatively low. Such artifacts include audible pre-echos resulting from quantization noise just before transients, 'warbling' or 'underwater' effects resulting from spectral holes, and alike. On the other hand, a parametric stereo coder may also result in audible artifacts if employed at low bitrates. Examples of such artifacts are a loss of 'width', audible crosstalk or instable sound source positioning. When these coding methods are combined in a single audio codec, it is likely that audible artifacts resulting from the core coder as well as parametric stereo will determine the overall perceived quality, and that the artifacts may 'add up' in some way when the perceived quality of the complete system is determined.

As discussed in Chapter 1, bitrate reduction in conventional lossy audio coders is obtained predominantly by exploiting the phenomenon of masking. Therefore, lossy audio coders rely on accurate and reliable masking models, which are often applied on individual channel signals in the case of a stereo or multi-channel signal. For a parametric stereo extended (mono) audio coder, however, the masking model is applied only once on a certain combination of the two input signals. This scheme has two implications with respect to masking phenomena.

The first implication relates to spatial unmasking of quantization noise. In stereo wave-form or transform coders, individual quantizers are applied on the two input signals or on linear combinations of the input signals. As a consequence, the injected quantization noise may exhibit different spatial properties than the audio signal itself. Due to binaural unmasking, the quantization noise may thus become audible, even if it is inaudible if presented monaurally. For tonal material, this unmasking effect (or BMLD, quantified as threshold difference between a binaural condition and a monaural reference condition) has shown to be relatively small (about 3 dB, see [127, 128]). However, it is expected that for broadband maskers, the unmasking effect is much more prominent. If one assumes an inter-aurally in-phase noise as masker, and a quantization noise which is either inter-aurally in-phase or inter-aurally uncorrelated, BMLDs are reported to be about 6 dB [72]. More recent data revealed BMLDs of 13 dB for this condition, based on a sensitivity of changes in the correlation of 0.045 [29]. To prevent these spatial unmasking effects of quantization noise, conventional stereo coders often apply some sort of spatial unmasking protection algorithm.

For a parametric stereo enhanced coder, on the other hand, there is only one waveform or transform quantizer, working on the mono (down-mix) signal. In the stereo reconstruction phase, both the quantization noise and the audio signal present in each frequency band will obey the same spatial properties. Since a difference in spatial characteristics of quantization noise and audio signal is a prerequisite for spatial unmasking, this effect is less likely to occur for parametric stereo enhanced coders than for conventional stereo coders.

5.3 Relation to BCC

The parametric stereo algorithm is based on identical principles as BCC and spatial audio coding techniques. The differences rely in certain implementation aspects and engineering choices.

- Parametric stereo supports dynamic segmentation of the incoming audio, to enable bit-rate scalability (by modifying the parameter update rate) and improve the coding efficiency. Signal parts that are reasonably stable in terms of spatial properties can be encoded with low parameter update rates (i.e., long analysis frames), while strong spatial dynamics can be accounted for as well by a temporarily higher update rate.

- A slightly different ICC, ICTD parameterization is used to allow parameterization of out-of-phase signals. The ICTD is replaced by an inter-channel phase difference (ICPD), which has the additional advantage of a lower estimation complexity in the encoder.

- A dedicated filterbank was designed that has a lower computational complexity and memory requirement than FFT-based methods, and which allows for efficient integration with aacPlus.

- An alternative decoder-side synthesis method has a low complexity, and allows for synthesis of out-of-phase output signals.

5.4 Parametric stereo encoder

5.4.1 Time/frequency decomposition

The encoder receives a stereo input signal pair $x_1(n)$, $x_2(n)$ with a sampling rate f_s. These input signals are decomposed in time/frequency tiles either using a STFT or by applying a filterbank. When using a STFT, time-domain segmentation and windowing is typically applied prior to transformation to the frequency domain. When a filterbank is applied, windowing and segmentation can be applied in the filterbank domain as well. If the input signal does not contain strong transients, the frame length (or parameter update rate) should match the lower bound of the measured time constants of the binaural auditory system (i.e., between 23 and 100 ms). Dynamic window switching is preferably used in the case of transients. The purpose of window switching is twofold. Firstly, to account for the precedence effect, which dictates that only the first 2 ms of a transient in a reverberant environment determine its perceived location. Secondly, to prevent ringing artifacts resulting from the frequency-dependent processing which is applied in otherwise relatively long segments. The window switching procedure, of which the essence is demonstrated in Figure 5.2, is controlled automatically by a transient detector.

If a transient is detected at a certain temporal position, a stop window of variable length is applied which just stops before the transient. The transient itself is captured using a very short window (of the order of a few milliseconds). A start window of variable length is subsequently applied to ensure segmentation at the same temporal grid as before the transient.

The frequency partitioning is organized in *parameter bands*. Each parameter band b ($b = 0, \ldots, B - 1$) represents a certain signal bandwidth for each time segment. The parameter bands closely mimic the critical band concept; at low frequencies, parameter bands are relatively narrow in bandwidth, while at high frequencies, the bandwidth of each parameter band increases.

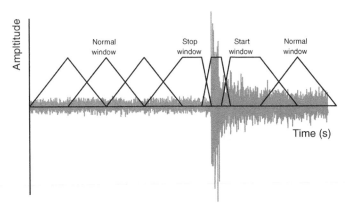

Figure 5.2 Schematic presentation of dynamic window switching in case of a transient. A stop window is placed just before the detected transient position. The transient itself is captured using a short window.

The frequency bands are formed in such a way that each band has a bandwidth, BW (in Hz), which is approximately equal to the equivalent rectangular bandwidth (ERB) [98], following:

$$BW = 24.7\,(0.00437f + 1) \tag{5.1}$$

with f the (center) frequency given in Hz. This process results in $B = 34$ parameter bands. The center frequencies of each parameter band vary between 28.7 Hz ($b = 0$) and 18.1 kHz ($b = 33$), assuming a sampling rate of 44.1 kHz. Additionally, two alternative parameter-band configurations are supported with a reduced number of parameter bands ($B = 20$ and $B = 10$). The bandwidths (as ratio of the full bandwidth, i.e., half the sampling frequency) of all parameter band configurations are given in Table 5.1.

5.4.2 Parameter extraction

The time/frequency decomposition process resulted in windowed sub-band or transformed signals $x_{1,m}(k)$ and $x_{2,m}(k)$ with m the sub-band index. The sub-band signals m are combined into parameter bands b, for which parameters are extracted. The first parameter is the inter-channel level difference ($\Delta L_{x_1 x_2, b}$), defined as the logarithm of the power ratio $p_{x_1, m \in m_b} / p_{x_2, m \in m_b}$ of corresponding parameter bands from the two input signals:

$$\Delta L_{x_1 x_2, b}(k_p) = 10 \log_{10} \frac{p_{x_1, m \in m_b}}{p_{x_2, m \in m_b}} \tag{5.2}$$

with $m \in m_b$ representing all sub-band signals corresponding to parameter band b and k_p the temporal parameter position corresponding to the current analysis frame. In the following, the parameter position k_p will not be shown, assuming that for each analysis segment, the parameter analysis procedure is repeated. Assuming complex-valued signals (for example resulting from an STFT or complex-valued filterbank), $\Delta L_{x_1 x_2, b}$ can be written as

$$\Delta L_{x_1 x_2, b} = 10 \log_{10} \frac{\sum_k \sum_{m \in m_b} x_{1,m}(k) x_{1,m}^*(k)}{\sum_k \sum_{m \in m_b} x_{2,m}(k) x_{2,m}^*(k)} \tag{5.3}$$

where (*) denotes complex conjugation.

The relative time difference between the channels is presented by a phase angle (inter-channel phase difference, ICPD). The use of ICPD over ICTD facilitates easier analysis and quantization than ICTD, given its limited range of 2π rad. Psychophysical evidence exists that the sensitivity to changes in ITD can be described quite accurately by a constant *phase* change (see Section 3.5.1) Hence the use of a phase parameter enables a single quantization strategy for all frequency bands. Another advantage of using ICPDs over ICTDs is that the estimation of ICTDs requires accurate unwrapping of bin-by-bin phase differences within each parameter band, which can be prone to errors. Last but not least, the ICPD parameter can be estimated directly in the filterbank or STFT domain, without the need of an additional transform to compute the time-domain cross-correlation function, hence resulting in a significant complexity reduction compared to ICTD estimation.

Table 5.1 Specification of parameter bandwidths for three spectral parameter resolution configurations. Parameter bandwidths are given as ratio of the full bandwidth of the filterbank.

Parameter band b	Bandwidth ($B = 34$)	Bandwidth ($B = 20$)	Bandwidth ($B = 10$)
0	1/384	1/256	1/128
1	1/384	1/256	1/128
2	1/384	1/256	1/64
3	1/384	1/256	1/64
4	1/384	1/128	2/64
5	1/384	1/128	2/64
6	1/256	1/128	2/64
7	1/256	1/128	6/64
8	1/256	1/64	9/64
9	1/256	1/64	40/64
10	1/128	1/64	n/a
11	1/128	1/64	n/a
12	1/128	1/64	n/a
13	1/128	1/64	n/a
14	1/128	2/64	n/a
15	1/128	3/64	n/a
16	1/64	4/64	n/a
17	1/64	5/64	n/a
18	1/64	12/64	n/a
19	1/64	29/64	n/a
20	1/64	n/a	n/a
21	1/64	n/a	n/a
22	2/64	n/a	n/a
23	2/64	n/a	n/a
24	2/64	n/a	n/a
25	2/64	n/a	n/a
26	2/64	n/a	n/a
27	3/64	n/a	n/a
28	3/64	n/a	n/a
29	3/64	n/a	n/a
30	3/64	n/a	n/a
31	4/64	n/a	n/a
32	4/64	n/a	n/a
33	23/64	n/a	n/a

The ICPD parameter $\phi_{x_1x_2,b}$ aims at optimal (in terms of correlation) phase alignment between the two input signals and is given by:

$$\phi_{x_1x_2,b} = \angle \left(\sum_k \sum_{m \in m_b} x_{1,m}(k) x_{2,m}^*(k) \right) \tag{5.4}$$

Using the ICPD as specified in Equation (5.4), (relative) delays between the input signals are represented as a constant phase difference in each parameter band, hence resulting in a piecewise constant phase characteristic which is a somewhat limited model for the delay. On the other hand, constant phase differences across the input signals are described accurately, which is in turn not possible if an ICTD parameter (i.e., a parameterized slope of phase with frequency) had been used. The (absence of) perceptual consequences by replacing time differences with parameter-band phase differences is discussed in Chapter 7.

The third parameter is the inter-channel coherence ($c_{x_1 x_2,b}$), which is, in this context, defined as the normalized cross-correlation coefficient after phase alignment according to the ICPD. The coherence $c_{x_1 x_2,b}$ is derived from the cross-spectrum by first applying the ICPD parameter to align the two input signals, followed by computing the cross-correlation of the phase-aligned signals:

$$c_{x_1 x_2,b} = \frac{\sum_k \sum_{m \in m_b} x_{1,m}(k) x_{2,m}^*(k) e^{j\phi_{x_1 x_2,b}}}{\sqrt{\left(\sum_k \sum_{m \in m_b} x_{1,m}(k) x_{1,m}^*(k)\right) \left(\sum_k \sum_{m \in m_b} x_{2,m}(k) x_{2,m}^*(k)\right)}} \tag{5.5}$$

which can also be written as

$$c_{x_1 x_2,b} = \frac{\left| \sum_k \sum_{m \in m_b} x_{1,m}(k) x_{2,m}^*(k) \right|}{\sqrt{\left(\sum_k \sum_{m \in m_b} x_{1,m}(k) x_{1,m}^*(k)\right) \left(\sum_k \sum_{m \in m_b} x_{2,m}(k) x_{2,m}^*(k)\right)}} \tag{5.6}$$

5.4.3 Down-mix

A suitable mono signal $s_{1,m}(k)$ is obtained by a linear combination of the input signals $x_{1,m}(k)$ and $x_{2,m}(k)$:

$$s_{1,m}(k) = w_{1,b}(k) x_{1,m}(k) + w_{2,b}(k) x_{2,m}(k) \tag{5.7}$$

where $w_{1,b}$ and $w_{2,b}$ are weights that determine the relative amount of $x_{1,m}(k)$ and $x_{2,m}(k)$ in the mono output signal. For example, if $w_{1,b} = w_{2,b} = 0.5$, the output will consist of the average of the two input signals. A down-mix that is created using fixed weights however bears the risk that the power of the down-mix signal strongly depends on the cross-correlation of the two input signals. To circumvent signal loss and signal coloration due to time- and frequency-dependent cross-correlations, the weights $w_{1,b}$ and $w_{2,b}$ are (1) complex-valued, to prevent phase cancellation, and (2) varying in magnitude across frequency or parameter bands (b), to ensure overall power preservation.

When the mono signal is generated, the last parameter that has to be extracted is computed. The ICPD parameter as described above specifies the *relative* phase difference between the stereo input signal (at the encoder) and the stereo output signals (at the decoder). Hence the ICPD does not indicate how the decoder should distribute these phase differences across the output channels. In other words, an ICPD parameter alone does not indicate whether a first signal is lagging the second signal, or vice versa. Thus, it is generally impossible to reconstruct the absolute phase for the stereo signal pair using only the relative phase difference. Absolute phase reconstruction is required to prevent signal cancellation in the applied overlap-add procedure in both the encoder as well as the decoder (see below). To signal the actual distribution of phase modifications, an overall channel phase difference (OCPD) is computed and transmitted. To be more specific, the decoder applies a phase modification equal to the OCPD to compute the first output signal, and applies a phase modification of the OCPD minus the ICPD to obtain the second output signal. Given this specification, the OCPD θ_b for parameter band b is computed as the average phase difference between $x_{1,m}(k)$ and $s_{1,m}(k)$, following

$$\theta_b = \angle \left(\sum_k \sum_{m \in m_b} x_{1,m}(k) s_{1,m}^*(k) \right) \tag{5.8}$$

Subsequently, the mono signal $s_{1,m}(k)$ is transformed to the time domain to result in the mono output signal.

5.4.4 Parameter quantization and coding

The ICLD, ICPD, OCPD and ICC parameters are quantized according to perceptual criteria. The quantization process aims at introducing quantization errors which are just inaudible. For the ICLD, this constraint requires a nonlinear quantizer, or nonlinearly spaced ICLD values given the fact that the sensitivity for changes in ICLD depends on the reference ICLD. The vector $\Delta \mathbf{L}$ contains the possible discrete ICLD values ΔL that are available for the quantizer. Each element in $\Delta \mathbf{L}$ represents a single quantization level for the ICLD parameter and is indicated by $\Delta L_q(i)$ $(i = (0, \ldots, 30))$:

$$\begin{aligned} \Delta \mathbf{L} = \ & \{\Delta L_q(0), \Delta L_q(1), \ldots, \Delta L_q(30)\} = \ldots \\ & \{-50, -45, -40, -35, -30, -25, -22, \ldots \\ & -19, -16, -13, -10, -8, -6, -4, -2, 0, \ldots \\ & 2, 4, 6, 8, 10, 13, 16, 19, 22, 25, 30, 35, 40, 45, 50\} \end{aligned} \tag{5.9}$$

The ICLD index for sub-band b, $i_{\Delta L_{x_1 x_2,b}}$, is then equal to

$$i_{\Delta L_{x_1 x_2,b}} = \arg \left(\min_i \left| \Delta L_{x_1 x_2,b} - \Delta L_q(i) \right| \right) \tag{5.10}$$

For the ICPD parameter, the vector ϕ represents the available quantized ICPD values ϕ_q:

$$\begin{aligned} \phi = \ & \{\phi_q(0), \phi_q(1), \ldots, \phi_q(7)\} = \ldots \\ & \{0, \pi/4, 2\pi/4, 3\pi/4, 4\pi/4, 5\pi/4, 6\pi/4, 7\pi/4\} \end{aligned} \tag{5.11}$$

This repertoire is in line with the finding that the human sensitivity to changes in timing differences at low frequencies can be described by a constant phase difference sensitivity. The ICPD index for sub-band b, $i_{\phi_{x_1 x_2,b}}$, is given by:

$$i_{\phi_{x_1 x_2,b}} = \text{mod}\left(\left\lfloor \frac{4\phi_{x_1 x_2,b}}{\pi} + \frac{1}{2}\right\rfloor, \Lambda_\phi\right) \qquad (5.12)$$

where mod(.) means the modulo operator, $\lfloor . \rfloor$ the floor function and Λ_ϕ the cardinality of the set of possible quantized ICPD values (i.e., the number of elements in ϕ). The OCPD (θ_b) is quantized using the same quantizer, resulting in i_{θ_b} according to

$$i_{\theta_b} = \text{mod}\left(\left\lfloor \frac{4\theta_b}{\pi} + \frac{1}{2}\right\rfloor, \Lambda_\theta\right) \qquad (5.13)$$

Finally, the repertoire for ICC, represented in the vector $\mathbf{c}$, is given by:

$$\begin{aligned} \mathbf{c} = \{c_q(0), c_q(1), \ldots, c_q(7)\} &= \ldots \\ \{1, 0.937, 0.84118, 0.60092, 0.36764, 0, -0.589, -1\} \end{aligned} \qquad (5.14)$$

This repertoire is based on just-noticeable differences in correlation reported by [63]. The coherence index $i_{c_{x_1 x_2,b}}$ for sub-band b is determined by

$$i_{c_{x_1 x_2,b}} = \text{arg}\left(\min_i |c_{x_1 x_2,b} - c_q(i)|\right) \qquad (5.15)$$

Thus, for each frame, B indices for the ICLD and ICC have to be transmitted. The ICPD and OCPD indices are not transmitted for sub-bands $b > 17$ for $B = 34$ (for $B = 20$ (or 10), only 11 (or 5) ICPD and OCPD parameters are transmitted). Hence the bandwidth for (interaural) phase reconstruction is limited to approximately 2400 Hz (assuming $f_s = 44100$ Hz), given the fact that the human auditory system is insensitive to fine-structure phase differences at high frequencies. ICTDs present in the high-frequency envelopes are supposed to be represented by the time-varying nature of ICLD parameters (hence discarding ICTDs presented in envelopes that fluctuate faster than the parameter update rate).

All parameters are transmitted differentially across time. The time-differential parameter indices are efficiently represented using entropy coding. The entropy per symbol, using modulo-differential coding and the resulting contribution to the overall bitrate (for $B = 34$ and a parameter update rate of 23 ms) are given in Table 5.2. These numbers were obtained by analysis of 80 different stereo audio recordings representing a large variety of material.

The total estimated parameter bitrate for the configuration as described above, excluding bitstream overhead, and averaged across a large amount of representative stereo material amounts to 7.7 kbit/s. If further parameter bitrate reduction is required, the following changes can be made.

- Reduction of the number of parameter bands (e.g., using 10 or 20 instead of 34). The parameter bitrate increases approximately linearly with the number of bands. This results in a bitrate of approximately 4.5 kbit/s for the 20 band case, assuming an update rate of 23 ms and including transmission of ICPD and OCPD parameters. Informal listening experiments showed that lowering the number of frequency bands below 10 results in severe degradation of the perceived spatial quality.

Table 5.2 Entropy per parameter symbol, number of symbols per second and bitrate for spatial parameters using 34 parameter bands and an update rate of 23 ms.

Parameter	Bits/symbol	symbols/second	bitrate (bits/s)
ICLD	1.94	1464	2840
ICPD	1.58	732	1157
OCPD	1.31	732	959
ICC	1.88	1464	2752
Total			7708

- No transmission of ICPD and OCPD parameters. As described above, the coherence is a measure of the difference between the input signals which cannot be accounted for by (sub-band) phase and level differences. A lower bitrate is obtained if the applied signal model does not incorporate phase differences. In that case, the normalized cross-correlation $\rho_{x_1x_2}$ is the relevant measure of differences between the input signals that cannot be accounted for by level differences. In other words, phase or time differences between the input signals are modeled as (additional) changes (decreases) in the coherence:

$$c_{x_1x_2,b} = \rho_{x_1x_2,b} \qquad (5.16)$$

The estimated cross-correlation value is then derived from the cross spectrum following:

$$\rho_{x_1x_2,b} = \frac{\mathrm{Re}\left(\sum_k \sum_{m\in m_b} x_{1,m}(k)x_{2,m}^*(k)\right)}{\sqrt{\left(\sum_k \sum_{m\in m_b} x_{1,m}(k)x_{1,m}^*(k)\right)\left(\sum_k \sum_{m\in m_b} x_{2,m}(k)x_{2,m}^*(k)\right)}} \qquad (5.17)$$

The associated average bit-rate reduction amounts to approximately 27% compared with parameter sets which do include the ICPD and OCPD values. Furthermore, the possible range of the ICC parameter now also includes negative values, which requires special attention at the decoder side (see also Section 5.5.3).

- Increasing the quantization errors of the parameters. The bitrate reduction is only marginal, given the fact that the distribution of time-differential parameters is very peaky.

- Decreasing the parameter update rate. The bitrate scales approximately linearly with the update rate.

In summary, the parameter bitrate can be scaled between approximately 8 kbit/s for maximum quality (using 34 parameter bands, an update rate of 23 ms and transmitting all relevant parameters) to about 1.5 kbit/s (using 20 parameter bands, an update rate of 46 ms, and no transmission of ICPD and OCPD parameters).

5.5 Parametric stereo decoder

The structure of the parametric stereo decoder is shown in Figure 5.3. The mono input signal s_1 is first processed by a hybrid QMF analysis filterbank. Subsequently, the hybrid QMF-domain signal $s_{1,m}$ is processed by a *decorrelator* (D) to result in a second signal $D(s_{1,m})$. These two input signals are processed by a matrix R_b. Finally, two hybrid QMF synthesis filterbanks generate the two domain–domain output signals $\hat{x}_1$ and $\hat{x}_2$. These separate stages will be explained in more detail in the following sections.

5.5.1 Analysis filterbank

The applied filterbank is a hybrid complex-modulated quadrature mirror filter bank (QMF) which is an extension to the filterbank as used in spectral band replication (SBR) techniques [68, 172, 279]. The hybrid QMF analysis filterbank consists of a cascade of two filterbanks. The structure is shown in the left panel of Figure 5.4.

The first filterbank (QMF analysis) is compatible with the filterbank as used in SBR. The sub-band signals which are generated by this first filterbank are obtained by convolving the input signal with a set of analysis filter impulse responses $G_{m_0}(n)$ given by

$$G_{m_0}(n) = g_0(n) \exp\left\{j\frac{\pi}{4M}(2m+1)(2n-1)\right\} \tag{5.18}$$

with $g_0(n)$ for $n = 0, \ldots, N_0 - 1$ the prototype window of the filter, $M_0 = 64$ the number of output channels, m_0 the sub-band index ($m_0 = 0, \ldots, M - 1$), and $N_0 = 640$ the filter

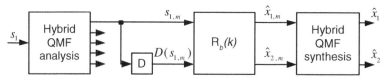

Figure 5.3   Structure of the QMF-based decoder. The signal is first fed through a hybrid QMF analysis filter bank. The filterbank output and a decorrelated version of each filterbank signal is subsequently fed into a matrixing stage $R_b(k)$. Finally, two hybrid QMF synthesis banks generate the two output signals.

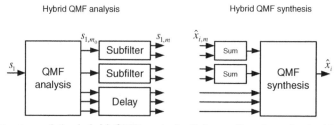

Figure 5.4   Structure of the hybrid QMF analysis (left panel) and synthesis (right panel) filter banks.

length. The filtered outputs are subsequently down sampled by a factor M_0, to result in a set of down-sampled QMF outputs (or sub-band signals) s_{1,m_0} (the equations given here are purely analytical; in practice the computational efficiency of the filter bank can be increased using decomposition methods).

$$s_{1,m_0}(k) = \big(s(n) * G_{m_0}(n)\big)(Mk) \tag{5.19}$$

The magnitude responses of the first four frequency bands ($m_0 = 0..3$) of the QMF analysis bank are illustrated in Figure 5.5.

The down-sampled sub-band signals s_{1,m_0} of the lowest QMF sub-bands are subsequently fed through a second complex-modulated filter bank (sub-filterbank) of order N_1 to further enhance the frequency resolution; the remaining sub-band signals are delayed by $N_1/2$ samples to compensate for the delay which is introduced by the sub-filterbank. The output of the hybrid (i.e., combined) filterbank is denoted by $s_{1,m}$, with m the index of the hybrid QMF bank. To allow easy identification of the two filterbanks and their outputs, the index m_0 of the first filterbank will be denoted 'sub-band index', and the index m_1 of the sub-filterbank is denoted 'sub-sub-band index'. The sub-filterbank has a filter order of $N_1 = 12$, and an impulse response $G_{m_1}(k)$ given by

$$G_{m_1}(k) = g_1(k) \exp\left\{ j \frac{2\pi}{M_1} (m_1 + \frac{1}{2})(k - \frac{N_1}{2}) \right\} \tag{5.20}$$

with $g_1(k)$ the prototype window, k the sub-band sample index, and M_1 the number of sub-sub-bands. Table 5.3 gives the number of sub-sub-bands $M_1(m_0)$ as a function of the QMF band m_0, for both the 34 and 20 analysis-band configurations. As an example the magnitude response of the 4-band sub-filterbank ($M_1 = 4$) is given in Figure 5.6.

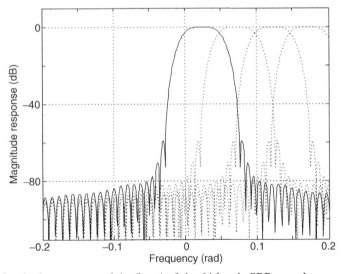

Figure 5.5 Magnitude responses of the first 4 of the 64 bands SBR complex-exponential modulated analysis filterbank. The magnitude for $m_0 = 0$ is highlighted.

Table 5.3 Specification of M_1 for the first 5 QMF sub-bands.

QMF sub-band (m_0)	$M_1(m_0)$ $(B = 34)$	$M_1(m_0)$ $(B = 20,10)$
0	12	8
1	8	4
2	4	4
3	4	1
4	4	1

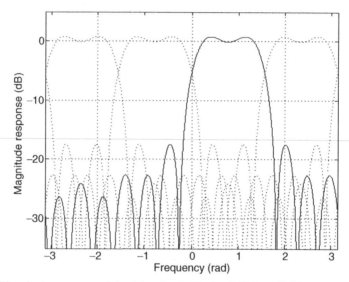

Figure 5.6 Magnitude response of the 4-band sub-filterbank ($M_1 = 4$). The response for $m_1 = 0$ is highlighted.

Obviously, due to the limited prototype length ($N_1 = 12$), the stop-band attenuation is only in the order of 20 dB.

As a result of this hybrid QMF filterbank structure, 91 (for $B = 34$) or 77 ($B = 20$ or 10) down-sampled filter outputs $s_{1,m}$ are available for further processing.

5.5.2 Decorrelation

In order to generate two output signals with a variable (i.e., parameter–dependent) coherence, a second signal is generated which has a similar spectral-temporal envelope as the mono input signal, but is incoherent from a fine-structure waveform point of view. This incoherent (or orthogonal) signal, $D(s_{1,m})$ is generated by the *decorrelator* (D), and is obtained by convolving the mono input signal with an all-pass filter. The decorrelation is performed in the QMF domain as shown in Figure 5.3. A very cost-effective decorrelation

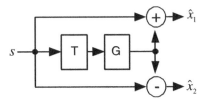

Figure 5.7   Structure of a Lauridsen decorrelator.

all-pass filter is obtained by a simple (frequency-dependent) delay. The combination of a delay and a (fixed) mixing matrix to produce two signals with a certain spatial diffuseness is known as a Lauridsen decorrelator [179]. The structure of a Lauridsen decorrelator is shown in Figure 5.7. The input signal s is delayed (T), attenuated (G) and added ($\hat{x}_1$) or subtracted ($\hat{x}_2$) from the input. The decorrelation is produced by complementary comb-filter peaks and troughs in the two output signals resulting from different signs in the combination of the direct and delayed signals.

The attenuation (G) associated with the delayed signal determines the coherence of the two output signals. A low value for G will result in a high coherence, while a value for G of +1 will result in fully decorrelated signals. The coherence $c_{x_1x_2}$ is given by:

$$c_{\hat{x}_1\hat{x}_2} = \frac{1 - G^2}{1 + G^2} \tag{5.21}$$

The delay T determines the spectral spacing of the comb filter. A longer delay results in a higher density of peaks and troughs. This can be observed from Figure 5.8. The top panel shows the two output spectra (represented by the solid and dashed lines, respectively) for a delay of $T = 5$ ms and a gain $G = 1$. For the lower panel, these parameters are 10 ms and 0.5, respectively. The longer delay clearly results in a more dense harmonic spectra, while the lower value of the gain results in a decrease in the depth of the spectral peaks and troughs.

This Lauridsen decorrelator works reasonably well provided that the delay is sufficiently long to result in multiple comb-filter peaks and troughs in each auditory filter. Due to the fact that the auditory filter bandwidth is larger at higher frequencies, the delay is preferably frequency dependent, being shorter at higher frequencies to prevent audible 'doubles' or 'echos'. A frequency-dependent delay has the additional advantage that it does not result in harmonic comb-filter effects in the output. To further increase the density of the comb-filter peaks and troughs, the parametric stereo decoder features a combination of a frequency-dependent delay, an IIR allpass filter to mimic properties of late reverberation, and a ducking mechanism to remove undesirable reverberation tails if the input signal exhibits a sudden strong decrease in energy. More information on the decorrelation all-pass filter can be found in [79].

One important aspect of the signal combination (or matrixing) that is applied in the Lauridsen decorrelator is that the single gain for mixing the delayed signal into both output channels with equal gain, but reversed phase is that this is only one of an infinite number possible realizations to realize a specific correlation. More specifically, the amount of decorrelation signal can be made smaller by introducing individual mixing control for

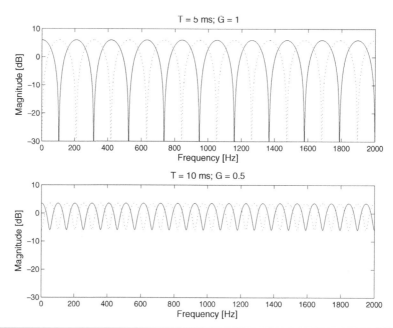

Figure 5.8   Spectra resulting from the Lauridsen decorrelator for $T = 5$ ms, $G = 1$ (top panel) and $T = 10$ ms, $G = 0.5$ (lower panel). The solid line represents output $\hat{x}_1$, the dotted line corresponds to $\hat{x}_2$.

each of the two output channels individually, resulting in a higher output quality. This will be outlined in the next section.

5.5.3 Matrixing

The matrixing stage $\mathbf{R}_b$ of the QMF-based spatial synthesis process performs a mixing and phase-adjustment process. For each sub-sub-band signal pair $s_{1,m}$, $D(s_{1,m})$, an output signal pair $\hat{x}_{1,m}$, $\hat{x}_{2,m}$ is generated by:

$$\begin{bmatrix} \hat{x}_{1,m}(k) \\ \hat{x}_{2,m}(k) \end{bmatrix} = \mathbf{R}_b(k) \begin{bmatrix} s_{1,m}(k) \\ D(s_{1,m})(k) \end{bmatrix} \qquad (5.22)$$

The mixing matrices $\mathbf{R}_b$ are determined by the following constraints:

1. The coherence of the two output signals must obey the transmitted ICC parameter.

2. The power ratio of the two output signals must obey the transmitted ICLD parameter.

3. The average energy of the two output signals must be equal to the energy of the mono input signal.

4. The average phase difference between the output signals must be equal to the transmitted ICPD value.

5. The average phase difference between $s_{1,m}$ and $x_{1,m}$ should be equal to the OCPD value.

Rotator 'A'

The parametric stereo decoder features two methods to re-create signals with the correct spatial properties. The first method, labelled 'A', aims at reconstruction of ICLDs and ICCs, without incorporation of phase differences (ICPD and OCPD). Due to the absence of ICPD parameters, the ICC parameter has a range of -1 to $+1$, and hence the mixing matrix should be designed in such a way that it can cope with this large correlation range in a robust way.

Synthesis of the correct ICC parameter using rotator 'A' can be understood by making the following matrix decomposition of the mixing matrix $\mathbf{R}_b$:

$$\mathbf{R}_b^A = \mathbf{Q}_b \mathbf{H}_b \tag{5.23}$$

with $\mathbf{H}_b$ a real-valued matrix to set the correct ICC parameter, and $\mathbf{Q}_b$ a diagonal, real-valued matrix that ensures the correct ICLD parameters by real-valued scaling. A suitable representation of the matrix $\mathbf{H}$ is given by:

$$\mathbf{H} = \begin{bmatrix} h_{11} & h_{12} \\ h_{21} & h_{22} \end{bmatrix} = \begin{bmatrix} \cos(\alpha + \beta) & \sin(\alpha + \beta) \\ \cos(-\alpha + \beta) & \sin(-\alpha + \beta) \end{bmatrix} \tag{5.24}$$

A visual interpretation of this matrix is shown in the left panel of Figure 5.9. The horizontal and vertical axes represent the two input signals $s_{1,m}$ and $D(s_{1,m})$. Each output signal ($\hat{x}_{1,m}$ and $\hat{x}_{2,m}$) is represented by a vector in the two-dimensional signal space. The two vectors have an angular difference of 2α and a mean (or common) rotation angle of β. The orthogonality property of the two input signals guarantees a fixed amount of energy that is represented by the length of the output vectors, independent of the two rotation angles α and β.

There exists a unique relation between the ICC parameter $c_{\hat{x}_1\hat{x}_2}$ and the rotation angle α which is given by:

$$\alpha = \frac{1}{2} \arccos\left(c_{\hat{x}_1\hat{x}_2}\right) \tag{5.25}$$

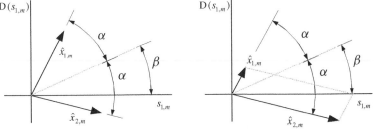

Figure 5.9 Visualization of rotator type 'A'.

Thus, the ICC value is independent of the overall rotation angle β. In other words, there exists an infinite number of solutions to combine two independent signals to create two output signals with a specified ICC value. In Figure 5.9, this is represented by the free choice of the angle β. The solution for β for rotator 'A' was chosen to maximize $r_{11} + r_{12}$, which also means minimization of $r_{21} + r_{22}$. Said differently, the mean output should predominantly consist of the mono input signal, while the decorrelator output signal should be minimized. This is visualized in the right panel of Figure 5.9. The two output signals are scaled according to the transmitted ICLD parameter. The sum vector of the two scaled output signals is placed exactly horizontally. This leads to the following solution for β_b:

$$\beta_b = \tan\left(\frac{\lambda_{2,b} - \lambda_{1,b}}{\lambda_{2,b} + \lambda_{1,b}} \arctan(\alpha_b)\right) \tag{5.26}$$

which can be approximated quite accurately by:

$$\beta_b = \tan\left(\frac{\lambda_{2,b} - \lambda_{1,b}}{\sqrt{2}}\alpha_b\right) \tag{5.27}$$

The variables $\lambda_{1,b}$ and $\lambda_{2,b}$ represent the relative amplitude of the two output signals with respect to the input and are given by:

$$\lambda_{1,b}^2 = 2\frac{10^{\left(\Delta L_{x_1 x_2,b}/10\right)}}{1 + 10^{\left(\Delta L_{x_1 x_2,b}/10\right)}} \tag{5.28}$$

$$\lambda_{2,b}^2 = 2\frac{1}{1 + 10^{\left(\Delta L_{x_1 x_2,b}/10\right)}} \tag{5.29}$$

The corresponding ICLD synthesis matrix $\mathbf{Q}_b$ is then given by

$$\mathbf{Q}_b = \begin{bmatrix} \lambda_{1,b} & 0 \\ 0 & \lambda_{2,b} \end{bmatrix} \tag{5.30}$$

Rotator 'B'

A second rotator type, rotator 'B', is based on a different decomposition of the mixing matrix. Rotator 'B' is applied when ICPD and OCPD parameters are present in the transmitted bitstream. For this rotator, the mixing matrix can be decomposed into three matrices $\mathbf{P}_b, \mathbf{A}_b, \mathbf{V}_b$:

$$\mathbf{R}_b^B = \sqrt{2}\,\mathbf{P}_b\mathbf{A}_b\mathbf{V}_b \tag{5.31}$$

The diagonal matrix $\mathbf{V}_b$ enables real-valued (relative) scaling of the two orthogonal signals $s_{1,m}$ and $D(s_{1,m})$. The matrix $\mathbf{A}_b$ is a real-valued rotation in the two-dimensional signal space, i.e., $\mathbf{A}_b^{-1} = \mathbf{A}_b^T$, and the diagonal matrix $\mathbf{P}_b$ enables modification of the complex-phase relationships between the output signals, hence $|p_{ij}| = 1$ for $i = j$ and 0 otherwise.

The solution for the matrix $\mathbf{P}b$ is given by:

$$\mathbf{P}_b = \begin{bmatrix} e^{j\theta_b} & 0 \\ 0 & e^{j\left(\theta_b - \phi_{x_1 x_2,b}\right)} \end{bmatrix} \qquad (5.32)$$

The matrices $\mathbf{A}_b$ and $\mathbf{V}_b$ can be interpreted as the eigenvector, eigenvalue decomposition of the covariance matrix of the (desired) output signals, assuming (optimum) phase alignment $(\mathbf{P}_b)$ prior to correlation. The solution for the eigenvectors and eigenvalues (maximizing the first eigenvalue v_{11} and hence minimizing the energy of the decorrelated signal) results from a singular value decomposition (SVD) of the covariance matrix. The matrices $\mathbf{A}_b$ and $\mathbf{V}_b$ are given by (see [134] for more details):

$$\mathbf{A}_b = \begin{bmatrix} \cos(\delta_b) & -\sin(\delta_b) \\ \sin(\delta_b) & \cos(\delta_b) \end{bmatrix} \qquad (5.33)$$

$$\mathbf{V}_b = \begin{bmatrix} \cos(v_b) & 0 \\ 0 & \sin(v_b) \end{bmatrix} \qquad (5.34)$$

with δ_b being a rotation angle in the two-dimensional signal space defined by $s_{1,m}$ and $D(s_{1,m})$, which is given by:

$$\delta_b = \begin{cases} \dfrac{\pi}{4} & \text{for } (c_{x_1 x_2,b}, \Delta L_{x_1 x_2,b}) \\ & = (0,0) \\[2ex] \text{mod}\left(\dfrac{1}{2}\arctan\left(\dfrac{2\frac{\lambda_{1,b}}{\lambda_{2,b}}c_{x_1 x_2,b}}{\frac{\lambda_{1,b}^2}{\lambda_{2,b}^2}-1}\right), \dfrac{\pi}{2}\right) & \text{otherwise} \end{cases} \qquad (5.35)$$

and v_b a parameter for relative scaling of $s_{1,m}$ and $D(s_{1,m})$ (i.e., the relation between the eigenvalues of the desired covariance matrix):

$$v_b = \arctan\sqrt{\dfrac{1 - \sqrt{\mu_b}}{1 + \sqrt{\mu_b}}} \qquad (5.36)$$

with

$$\mu_b = 1 + \dfrac{\left(4c_{x_1 x_2,b}^2 - 4\right)\lambda_{1,b}^2\lambda_{2,b}^2}{\lambda_{1,b}^4 + \lambda_{2,b}^4} \qquad (5.37)$$

A visualization of this rotator is shown in Figure 5.10. The two (desired) output signals are shown along the horizontal and vertical axis. All sample pairs $\hat{x}_{m,1}, \hat{x}_{m,2}$ can be represented as points in the two-dimensional space (plane) shown in Figure 5.10. A large set of points form a oval-like shape. The size (in the vertical and horizontal direction) is determined by the variance (or power) of $\hat{x}_{m,1}$ and $\hat{x}_{m,2}$, respectively. Its orientation (the angle δ) is dependent on the coherence between $\hat{x}_{m,1}$ and $\hat{x}_{m,2}$. The oval shape is decomposed into a dominant signal $s_{m,1}$ (the dominant component, having maximum

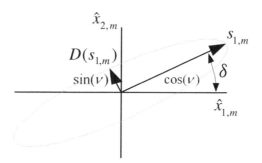

Figure 5.10   Visualization of rotator 'B'.

power, which is equal to the square of the largest eigenvalue) and a residual signal $D(s_{m,1})$ (the residual component, with a power equal to the square of the smallest eigenvalue).

It should be noted that a two-dimensional eigenvector problem has in principle four possible solutions: each eigenvector, which is represented as columns in the matrix **A**, may be multiplied with a factor -1. The modulo operator in Equation (5.35) ensures that the first eigenvector is always positioned in the first quadrant. This is very important to prevent sign changes of signal components across frames (which would lead to signal 'dropouts'). However, this technique only works under the constraint of $c_{x_1 x_2, b} > 0$, which is guaranteed if phase-alignment is applied. If no ICPD/OCPD parameters are transmitted, however, the ICC parameters may become negative, which hence requires rotator type 'A' since this rotator provides a stable solution for negative ICCs.

5.5.4 Interpolation

For each transmitted parameter the mixing matrix $\mathbf{R}_b$ is determined as described previously. However these matrices correspond in most cases to a single time instance, which depends on the segmentation and windowing procedure of the encoder. The sample index k at which a parameter set is valid is denoted by k_p, which is referred to as the *parameter position*. The parameter positions are transmitted from encoder to decoder along with the corresponding parameters themselves. For that particular QMF sample index $(k = k_p)$, the mixing matrices $\mathbf{R}_b$ are determined as described previously. For QMF sample indices (k) in between parameter positions, the mixing matrices are interpolated linearly (i.e., its real and imaginary parts are interpolated individually). The interpolation of mixing matrices has the advantage that the decoder can process each 'slot' of hybrid QMF samples (i.e., one sample from each sub-band) one by one, without the need of storing a whole frame of sub-band samples in memory. This results in a significant memory reduction compared to frame-based synthesis methods.

The decoder-side interpolation is outlined in Figure 5.11. The top panel shows encoder-side segmentation and windowing for a signal containing a transient.

The decoder is organized in frames. Each frame may contain one or more parameter sets, including corresponding parameter positions within the frame. By default, parameters are valid at the end of a frame, and for temporal positions for which no parameters are specified, the mixing matrices are interpolated linearly. This is indicated for the first

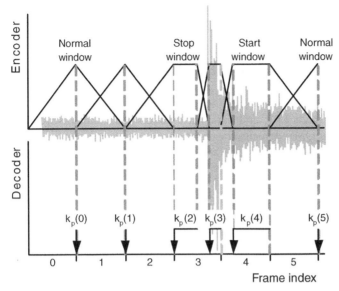

Figure 5.11 Encoder-side segmentation and windowing (top panel) and corresponding decoder–side parameter positions (lower panel).

three frames $(0 \ldots 2)$. For these frames, 'normal' analysis windows are applied in the encoder. The corresponding decoder parameter positions (k_p) coincide with the maximum (temporal center) of the analysis window. Frame '3' contains a transient. The encoder applies a stop window that ends just before the transient. The corresponding decoder-side interpolation first 'holds' the parameters from frame '2'. A new parameter set and corresponding position $(k_p(3))$ is applied at a position that corresponds to the encoder-side transient window. These parameters are also valid for an extended range of sample indexes, corresponding to the size of the plateau of the encoder-side transient window. Frame '4' again has a specified parameter position because this is the only parameter set transmitted for frame '4', these parameters are valid until the end of the frame (since no interpolation can be performed because the parameters from frame '5' are not known when processing frame '4'). Finally, frame '5' is processed with default parameter positions at the end of the frame.

5.5.5 Synthesis filterbanks

The mixing process is followed by a pair of hybrid QMF synthesis filterbanks (one for each output channel), which also consists of two stages (see Figure 5.4, right panel). The first stage comprises summation of the sub-sub-bands m_1 which stem from the same sub-band m_0:

$$\begin{cases} \hat{x}_{1,m_0} = \sum_0^{M_1(m_0)-1} \hat{x}_{1,m_1} \\ \hat{x}_{2,m_0} = \sum_0^{M_1(m_0)-1} \hat{x}_{2,m_1} \end{cases} \qquad (5.38)$$

Finally, up-sampling and convolution with synthesis filters (which are similar to the QMF analysis filters as specified by Equation 5.18) results in the final stereo output signal.

5.5.6 Parametric stereo in enhanced aacPlus

As described in the previous sections, enhanced aacPlus (or aacPlus v2) combines an AAC core codec with SBR and parametric stereo (PS). Since SBR and PS operate in virtually the same QMF domain, these parametric extensions can be combined in a very effective way, resulting in a significant complexity reduction compared to an operation mode where SBR and PS are both used independently. The structure of the enhanced aacPlus decoder is shown in Figure 5.12.

The incoming bitstream is demultiplexed into a (mono) AAC bitstream, SBR parameters and PS parameters. Subsequently, the AAC decoder generates a mono output signal of limited bandwidth. This signal is processed by a QMF analysis filterbank. The number of sub-bands amounts to 32, because of the limited signal bandwidth of the AAC decoder. The SBR process subsequently generates the upper half of the signal bandwidth, resulting in 64 sub-band signals. The delay of the SBR process amounts 6 QMF samples, which is exactly identical to the delay caused by the sub-filterbank of the hybrid QMF filter bank required for PS. In other words, the upper part (QMF band 32 to 63) of the signal (generated by SBR) can serve as direct input to the PS synthesis stage. The lower QMF bands (0 to 31) should only be processed by the sub-filterbank. In fact, the sub-filterbank only requires filtering in the first few QMF bands (see Table 5.3), while the remaining bands of the lower half of the bandwidth are simply delayed.

The resulting full-bandwidth hybrid QMF signal is processed by the parametric stereo decoder to generate two, full-bandwidth, hybrid QMF domain output signals. Finally, two hybrid QMF synthesis filter banks result in the two time-domain output signals.

To verify the compression gain resulting from parametric stereo, listening tests were conduced. The listening tests were carried out in two laboratories (indicated by black and gray in Figure 5.13) with 8 or 10 subjects, respectively. 10 test items from the MPEG-4 aacPlus stereo verification test [141] were used as test material. The coded excerpts included aacPlus v1 using normal stereo coding operating at 24 and 32 kbps, as well as aacPlus v2 operating at 24 kbps total. Two lowpass anchors (with cutoff frequencies of 3.5 and 7 kHz) and a hidden reference were also included in the test. Subjects had to rate the perceptual quality of each codec in a MUSHRA test [148]. The tests were performed in a sound-proof listening room using headphones. Figure 5.13 shows subjective listening test results averaged across listeners per test lab. The horizontal axes indicates the audio

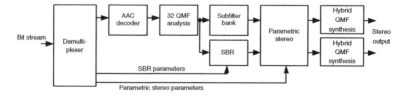

Figure 5.12   Structure of enhanced aacPlus (aacPlus v2).

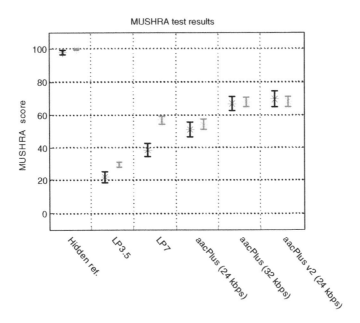

Figure 5.13 Subjective listening test results comparing aacPlus with aacPlus v2.

codec; the vertical axis shows the corresponding MUSHRA score. Error bars denote the 95% confidence intervals.

At both test sites, it was found that aacPlus v2 at 24 kbps achieves an average subjective quality that is equal to aacPlus v1 at 32 kbps, and is significantly better than aacPlus v1 at 24 kbps. In other words, parametric stereo results in an additional compression gain of 33% at bitrates in the range of 24–32 kb/s.

5.6 Conclusions

The parametric stereo coder has been described and its embedding in aacPlus v2 has been outlined. The coder employs analysis and synthesis of ICLD, ICPD and ICC parameters and is quite flexible in terms of parameter update rate, temporal framing and spectral resolution. The implementation of parametric stereo in the (hybrid) QMF domain enables an efficient combination of parametric stereo and SBR.

Listening tests have revealed that parametric stereo in aacPlus v2 provides a bitrate gain of about 33% over (stereo) aacPlus v1, enabling high-quality stereo audio transmission at bitrates of only 24 to 32 kb/s. This unsurpassed compression efficiency makes aacPlus v2 especially attractive for mobile applications, where bandwidth is relatively scarce and expensive.

6

MPEG Surround

6.1 Introduction

Approximately half a century after the introduction of two-channel stereophony, multi-channel sound is now on its way into consumers homes as the next step toward more realistic audio reproduction. Initially, multi-channel audio was predominantly present in the movie domain on consumer media (DVD for example). The widespread availability of movie material with multi-channel sound tracks led to a fast penetration of multi-channel playback devices in consumers homes. Recently, probably in part due to the increased popularity of multi-channel movie material, the demand for a compelling surround experience has extended to the audio-only market as well (such as SACD and DVD-audio).

In contrast, the traditional broadcast services (such as radio and television) are still operating in stereo mode due to bandwidth and compatibility constraints. In audio transmission systems, the required bandwidth (or amount of information) of a six-channel broadcast would require approximately three times as much bandwidth as a conventional stereo broadcast. In many cases, this increased amount of information is undesirable or unavailable. Even if the increased bandwidth were available, the upgrade process of a stereo service to multi-channel audio should ensure that existing stereo receivers will still operate as before. With the existing technology, this means an even larger increase in bandwidth for simulcast of stereo and multi-channel audio.

In this chapter, the recently finalized ISO/MPEG standard, 'MPEG Surround', is outlined. Given the complexity and the large variety of features of this standard, the focus is on describing the basic processing stages and their relations, rather than giving a detailed system description.

The MPEG Surround standard emerged from activities of the MPEG Audio standardization group. In March 2004, MPEG issued a call for proposals (CfP) requesting for technology in the field of spatial audio coding [142]. In response to this CfP various companies responded with a total of four submissions. The subjective evaluation of these submissions was concluded in October 2004. The test results revealed that there were two out of the four submissions that showed complementary performance. The proponents of both systems decided to cooperate and define a single system, to combine the best of both propositions. Beginning 2005, this resulted in Reference Model 0 (RM0), the starting point

Spatial Audio Processing: MPEG Surround and Other Applications   Jeroen Breebaart and Christof Faller
© 2007 John Wiley & Sons, Ltd

for the collaborative phase within the MPEG Audio group. Numerous core experiments have been conducted by various companies in order to improve and extend the MPEG Surround system, including a low-complexity mode and a dedicated binaural decoding mode to simulate a virtual multi-channel loudspeaker setup over stereo headphones. The standard specification of MPEG Surround [138] was finalized in July 2006.

6.2 Spatial audio coding

6.2.1 Concept

The concept of spatial audio coding as employed in the MPEG Surround standard [138] is outlined in Figure 6.1. A multi-channel input signal is converted to a down-mix by an MPEG Surround *encoder*. Typically, the down-mix is a mono or a stereo signal, but more down-mix channels are also supported (for example a 5.1 down-mix from a 7.1 input channel configuration). The perceptually relevant spatial properties of the original input signals that are lost by the down-mix process are captured in a *spatial parameter* bitstream. The down-mix can subsequently be encoded with an existing compression technology. In the last encoder step, the spatial parameters are combined with the down-mix bitstream by a multiplexer to form the output bitstream. Preferably, the parameters are stored in an ancillary data portion of the down-mix bitstream to ensure backward compatibility.

The right panel of Figure 6.1 outlines the MPEG Surround decoding process. In a first stage, the transmitted bitstream is split into a down-mix bitstream and a spatial parameter stream. The down-mix bitstream is decoded using a legacy decoder. Finally, the multi-channel output is constructed by an MPEG Surround decoder based on the transmitted spatial parameters.

The use of an MPEG Surround encoder as a pre-processor for a conventional (legacy) codec (and a corresponding post-processor in the decoder) has important advantages over existing multi-channel compression methods.

- The parametric representation of spatial properties results in a significant compression gain over conventional multi-channel audio codecs, as will be shown in Section 6.5.

- The use of a legacy codec with an additional spatial parameter stream allows for backward compatibility with existing compression schemes and broadcast services.

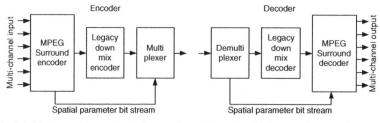

Figure 6.1 Multi-channel encoder (left panel) and decoder (right panel) according to the spatial audio coding concept. Reproduced by permission of the Audio Engineering Society, Inc, New York, USA.

The MPEG Surround coder inherited many of the aspects of parametric stereo, such as the support for dynamic segmentation, the dedicated filterbank and different parameter frequency resolutions. Several new features were developed as well:

- The spatial parameterization enables novel techniques to process or modify certain aspects of a down mix. Examples are matrixed-surround compatible down-mixes, support for so-called *artistic down-mixes* or the generation of a 3D/binaural signal to evoke a multi-channel experience over legacy headphones.

- The channel configuration at the spatial encoder can be different from the channel configuration of the spatial decoder without the need of full multi-channel decoding as intermediate step. For example, a decoder may directly render an accurate four-channel representation from a 5.1 signal configuration without having to decode all 5.1 channels first.

- To overcome limitations of a parametric model, 'residual coding' was introduced to enable MPEG Surround to support a higher quality, approaching transparency.

- A enhanced matrix mode allows for upmixing conventional stereo content to high-quality multi-channel signals.

6.2.2 Elementary building blocks

The MPEG Surround spatial coder structure is composed of a limited set of elementary building blocks. Each elementary building block is characterized by a set of input signals, a set of output signals, and a parameter interface. The generic elementary building block is shown in Figure 6.2. An elementary building block can have up to three input and output signals (as shown on the left and right side, respectively), as well as an input or output for (sets of) spatial parameters.

Different realizations of elementary building blocks serve different purposes in the spatial coding process. For example, a first type of building block may decrease the number of audio channels by means of spatial parameterization. Hence, if such a block is applied at the encoder side, the block will have fewer output channels than input channels, and has a parameter *output*. The corresponding block at the decoder side, however, has a parameter *input* and more output channels than input channels. The encoder and decoder representations of such an *encoding/decoding* block are shown in the top left and top right panels of Figure 6.3, respectively. Two different realizations of the encoding/decoding blocks exist. The first realization is a block that describes two signals as one down-mix signal and parameters. The corresponding encoding block is referred to as two-to-one

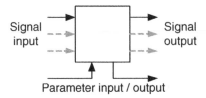

Signal input   Signal output

Parameter input / output

Figure 6.2 Generic elementary building block for the MPEG Surround coding process. Reproduced by permission of the Audio Engineering Society, Inc, New York, USA.

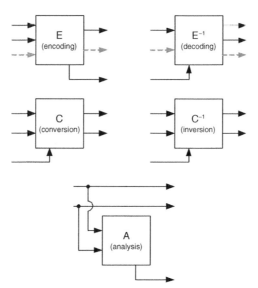

Figure 6.3   Elementary building blocks for MPEG Surround coding process. Reproduced by permission of the Audio Engineering Society, Inc, New York, USA.

(TTO), while the decoding block is termed one-to-two (OTT). In essence, these blocks are similar to a *parametric stereo* encoder/decoder [47, 79, 214, 217, 234, 235]. The second realization is a so-called three-to-two (TTT) encoding block, which generates two output signals and parameters from three input signals. The corresponding two-to-three decoding block generates three signals from a stereo input accompanied by parameters.

A second type of building block is referred to as signal *converter*. For example, a stereo input signal may be converted into a stereo output signal that has different spatial properties, and of which the processing is controlled by parameters. This is shown by the left-middle panel of Figure 6.3. The corresponding decoder-side operation (as shown in the right-middle panel) *inverts* the processing that is applied at the encoder to retrieve the original (unmodified) stereo input signal. Examples of signal converters are the conversion from conventional stereo to matrixed surround compatible stereo or to 3D/binaural stereo for playback over headphones.

The third type of building block is an *analysis* block. This type generates parameters from a signal stream without modifying the actual signals or signal configuration. This block, that can be applied at both the spatial encoder as well as the decoder side, is shown in the bottom panel of Figure 6.3.

6.3 MPEG Surround encoder

6.3.1 Structure

The structure of the MPEG Surround encoder is shown in Figure 6.4. A multi-channel input signal is first processed by a channel-dependent pre-gain. These gains enable adjustment of the level of certain channels (for example LFE and surround) within the transmitted down

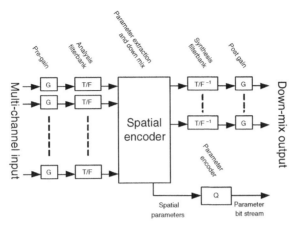

Figure 6.4 Structure of the MPEG Surround encoder. Reproduced by permission of the Audio Engineering Society, Inc, New York, USA.

mix. Subsequently, the input signals are decomposed into time/frequency tiles using an analysis filter bank. A spatial encoder generates a down-mix signal and (encoded) spatial parameters for each time/frequency tile. These parameters are quantized and encoded into a parameter bitstream by a parameter encoder (Q). The down-mix is converted to the time domain using a synthesis filterbank. Finally, a post-gain is applied to control the overall signal level of the down-mix.

6.3.2 Pre- and post-gains

In the process of down-mixing a multi-channel signal to a stereo signal, it is often desirable to have unequal weights for the different input channels. For example, the surround channels are often attenuated by 3 dB prior to the actual down-mix process. MPEG Surround supports user-controllable pre-gains between 0 and −6 dB, in steps of 1.5 dB. For the LFE, these weights are adjustable between 0 and −20 dB in steps of 5 dB.

The level of the generated down-mix can also be controlled using (post-encoder) gains to prevent clipping in the digital signal domain. The down-mix can be attenuated between 0 and −12 dB in steps of 1.5 dB.

The applied pre- and post-gain factors are signaled in the MPEG Surround bitstream to enable their inverse scaling at the decoder side.

6.3.3 Time–frequency decomposition

Filterbank

The applied filter bank is a hybrid complex-modulated quadrature mirror filterbank (QMF) that has the same structure as the filterbank applied in Parametric Stereo (see Chapter 5). Table 6.1 gives the number of sub-sub-bands $M_1(m_0)$ as a function of the QMF band m_0 that are used for MPEG Surround.

Table 6.1 Specification of M_1 and the resulting number of output channels for the first 3 QMF sub-bands.

QMF sub-band (m_0)	$M_1(m_0)$
0	8
1	4
2	4

The resulting sub-sub-band signals are grouped into so-called parameter bands which share common spatial parameters. Each parameter band comprises one or a set of adjacent sub-sub-bands to form the corresponding time/frequency tiles for which spatial parameters are estimated. For the highest frequency resolution supported by MPEG Surround, the number of parameter bands amounts to 28. Bitrate/quality trade-offs are supported by coarser frequency resolutions, resulting in different combinations of sub-sub-band signals into respective parameter bands. The following alternative number of parameter bands are supported: 4, 5, 7, 10, 14, and 20.

The sub-band signals are split into (time) segments in a similar way as described for Parametric Stereo (cf. Chapter 5) using dynamic segmentation that is adapted to the input signals.

6.3.4 Spatial encoder

Tree structures

The elementary building blocks (as described in Section 6.2.2) are combined to form a spatial coding *tree*. Depending on the number of (desired) input and output channels, and additional features that are employed, different tree structures may be constructed. The most common tree structures for 5.1-channel input will be outlined below. First, two tree structures for a mono down mix will be described, followed by the preferred tree structure for a stereo down-mix.

The first tree structure supports a mono down-mix and is outlined in the left panel of Figure 6.5. The six input channels, left front, right front, left surround, right surround, center and low-frequency enhancement, labeled l_f, r_f, l_s, r_s, c and *LFE*, respectively, are combined pairwise using encoding blocks (TTO type) until a mono down-mix is obtained. Each TTO block produces a set of parameters P. As a first step, the two front channels (l_f, r_f) are combined into an TTO encoding block E₃, resulting in parameters P_3. Similarly, the pairs c, *LFE* and l_s, r_s are combined by TTO encoding block E₄ and E₂, respectively. Subsequently, the combination of l_f, r_f on the one hand, and c, *LFE* on the other hand are combined using TTO encoding block E₁ to form a 'front' channel f. Finally, this front channel is merged with the common surround channel in encoding block E₀ to result in a mono output s.

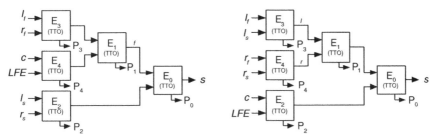

Figure 6.5 Tree configurations for a mono down-mix. Reproduced by permission of the Audio Engineering Society, Inc, New York, USA.

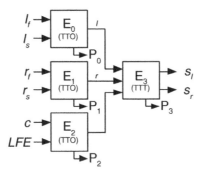

Figure 6.6 Preferred tree configuration for a stereo down-mix. Reproduced by permission of the Audio Engineering Society, Inc, New York, USA.

One of the advantages of this structure is its support for configurations with only one surround channel. In that case, l_s and r_s are identical and hence the corresponding TTO block can be omitted (i.e., the tree can be pruned).

The second tree structure for 5.1 input combined with a mono down mix is shown in the right panel of Figure 6.5. In this configuration, the l_f and the l_s channel are first combined into a common left channel (l) using an TTO encoding block E_3. The same process is repeated for the r_f and the r_s channel (E_4). The resulting common left and the common right channels are then combined in E_1, and finally merged (E_0) with the combination of the center and LFE channel (E_2). The advantage of this scheme is that a front-only channel configuration (i.e., only comprising l, r and c) is simply obtained by pruning the tree.

For a stereo down-mix, the preferred tree configuration is given in Figure 6.6. As for the second mono-based tree, this tree also starts by generation of common left and right channels, and a combined center/LFE channel. These three signals are combined into a stereo output signal s_l, s_r using a TTT encoding block (E_3).

TTO encoding block

The TTO encoding block transforms two input channels $x_{1,m}$, $x_{2,m}$ into one mono output channel $s_{1,m}$ plus spatial parameters. Its concept is identical to a parametric stereo encoder (see [47, 79, 214, 217, 234, 235]). For each parameter band, two spatial parameters are extracted. The first comprises the ICLD ($\Delta L_{x_1 x_2, b}$) between the two input channels for each parameter band b:

$$\Delta L_{x_1 x_2, b} = 10 \log_{10} \frac{p_{x_1, b}}{p_{x_2, b}} \tag{6.1}$$

with $p_{x_i, b}$ the power of signal x_i in parameter band b:

$$p_{x_i, b} = \sum_k \sum_{m=m_b}^{m_{b+1}-1} x_{i,m}(k) x_{i,m}^*(k) \tag{6.2}$$

where m_b represents the hybrid start band of parameter band b (sub-sub-band sample index) and k the time slot of the windowed segment. The second parameter is the inter-channel correlation ($\rho_{x_1 x_2, b}$):

$$\rho_{x_1 x_2, b} = \mathrm{Re} \left\{ \frac{\sum_k \sum_{m=m_b}^{m_{b+1}-1} x_{1,m}(k) x_{2,m}^*(k)}{\sqrt{p_{x_1, b} p_{x_2, b}}} \right\} \tag{6.3}$$

The mono down-mix $s_{1,m}$ comprises a linear combination of the two input signals. The associated down-mix weights for each input channel are determined based on the following decomposition of the two input signals:

$$x_{1,m}(k) = \psi_{1,b} s_{1,m}(k) + d_{1,m}(k) \tag{6.4}$$

$$x_{2,m}(k) = \psi_{2,b} s_{1,m}(k) - d_{1,m}(k) \tag{6.5}$$

Hence, the two input signals are described by a common component $s_{1,m}$ which may have a different contribution to $x_{1,m}$ and $x_{2,m}$ (represented by the coefficients $\psi_{i,b}$), and an out-of-phase component $d_{1,m}$ which is, except for the sign, identical in both channels. Furthermore, energy preservation is imposed by demanding the signal $s_{1,m}$ to have an energy that is equal to the sum of the energies of both input signals. The signal $s_{1,m}$, the desired mono down-mix signal, is given by:

$$s_{1,m}(k) = \frac{x_{1,m}(k) + x_{2,m}(k)}{\psi_{1,b} + \psi_{2,b}} \tag{6.6}$$

The energy preservation constraint results in:

$$(\psi_{1,b} + \psi_{2,b})^2 = \frac{p_{x_1,b} + p_{x_2,b} + 2\rho_{x_1 x_2, b} \sqrt{p_{x_1,b} p_{x_2,b}}}{p_{x_1,b} + p_{x_2,b}} \tag{6.7}$$

The signal $d_{1,m}$ is the *residual* signal. This signal is either discarded at the encoder side (in the case of a fully parametric description of the input signals, where *synthetic* residual signals are used at the decoder side) or can be transmitted to enable full waveform reconstruction at the decoder side. A hybrid approach is also facilitated: a specified low-frequency part of the residual signals can be selected for transmission, while for the remaining signal bandwidth, the residual signals are substituted by synthetic signals at the decoder. This option makes the system very flexible in terms of quality/bitrate trade-offs.

TTT encoding block using prediction mode

The TTT encoding block has three inputs (x_l, x_r, x_c), two down-mix outputs (s_l, s_r) and an auxiliary signal (s_c). The two outputs and the auxiliary signal form a linear combination of the input signals according to:

$$
\begin{bmatrix} s_{l,m}(k) \\ s_{r,m}(k) \\ s_{c,m}(k) \end{bmatrix} = \begin{bmatrix} 1 & 0 & 1 \\ 0 & 1 & 1 \\ 1 & 1 & -1 \end{bmatrix} \begin{bmatrix} x_{l,m}(k) \\ x_{r,m}(k) \\ x_{c,m}(k)\frac{1}{2}\sqrt{2} \end{bmatrix}
\tag{6.8}
$$

The center signal x_c is attenuated by 3 dB to ensure preservation of the center-channel power in the down-mix. The auxiliary output signal, s_c, which has orthogonal down-mix weights, would in principle allow full reconstruction of the three input signals by applying the inverse of the down-mix matrix as up-mix matrix. This would result in:

$$
\begin{bmatrix} \hat{x}_{l,m}(k) \\ \hat{x}_{r,m}(k) \\ \hat{x}_{c,m}(k)\frac{1}{2}\sqrt{2} \end{bmatrix} = \frac{1}{3} \begin{bmatrix} 2 & -1 & 1 \\ -1 & 2 & 1 \\ 1 & 1 & -1 \end{bmatrix} \begin{bmatrix} s_{l,m}(k) \\ s_{r,m}(k) \\ s_{c,m}(k) \end{bmatrix}
\tag{6.9}
$$

However, this third signal s_c is discarded at the encoder side and replaced by two prediction coefficients that enable an estimation $\hat{s}_c$ from the two down-mix channels s_l, s_r:

$$
\hat{s}_{c,m}(k) = \gamma_{1,b} s_{l,m}(k) + \gamma_{2,b} s_{r,m}(k)
\tag{6.10}
$$

with $\gamma_{1,b}, \gamma_{2,b}$ two channel prediction coefficients (CPCs) for each parameter band b. The prediction error $d_{1,m}$

$$
d_{1,m}(k) = s_{c,m}(k) - \hat{s}_{c,m}(k)
\tag{6.11}
$$

may be either transmitted or discarded, depending on the desired quality/bitrate trade-off. If the residual signal $d_{1,m}$ is discarded, the corresponding energy loss is described by an ICC parameter ρ_b:

$$
\rho_b^2 = 1 - \frac{p_{d_{1,m},b}}{p_{s_l,b} + p_{s_r,b}^2 + \frac{1}{2}p_{s_c,b}}
\tag{6.12}
$$

This ICC parameter ρ describes the ratio between the sum of the energies of the reconstructed $\hat{x}_l$, $\hat{x}_r$ and $\hat{x}_c$ signals using the prediction $\hat{s}_c$ and the sum of the energies of the original input signals:

$$
\rho_b^2 = \frac{p_{\hat{x}_l,b} + p_{\hat{x}_r,b} + \frac{1}{2}p_{\hat{x}_c,b}}{p_{x_l,b} + p_{x_r,b} + \frac{1}{2}p_{x_c,b}}
\tag{6.13}
$$

If the prediction error $d_{1,m}$ is zero, the ICC parameter will be exactly $+1$. Lower values indicate a prediction error (i.e., a prediction loss).

TTT encoding block using energy mode

The predictive mode for the TTT encoding block requires a reliable estimate of the signal s_c at the decoder side. If waveform accuracy cannot be guaranteed (for example in the high-frequency range of an audio coder employing SBR), a different TTT encoding mode is supplied which does not rely on specific waveforms, but only describes the relative energy distribution of the three input signals using two ICLD parameters:

$$\Delta L_{1,b} = 10 \log_{10} \frac{p_{x_l,b} + p_{x_r,b}}{\frac{1}{2} p_{x_c,b}} \qquad (6.14)$$

$$\Delta L_{2,b} = 10 \log_{10} \frac{p_{x_l,b}}{p_{x_r,b}} \qquad (6.15)$$

The prediction and energy mode can be used independently in different bands. In that case, parameter bands of a specified (lower) frequency range applies prediction parameters, while the remaining (upper) parameter bands apply the energy mode.

MTX conversion block

Matrixed surround (MTX) refers to a method to create a *pseudo* surround experience based on a stereo down-mix with specific down-mix properties. In conventional matrixed surround systems, the down-mix $(s_{l_{MTX}}, s_{r_{MTX}})$ is created such that signals of the surround channels are down-mixed in anti-phase. The anti-phase relationship of the surround channels in the down-mix enables a matrixed surround decoder to control its front/surround panning. The drawback of this static down-mix matrix is that it is impossible to retrieve the original input channels, nor is it possible to reconstruct a conventional stereo down-mix from the matrixed surround compatible down-mix. In MPEG Surround, however, a matrixed surround mode is supplied for compatibility with legacy matrixed surround devices and hence this option must not have any negative impact on any MPEG Surround operation. Therefore, the approach of MPEG Surround to create a matrixed surround compatible down-mix is different from the static down-mix approach of conventional matrixed surround encoders. A conversion from a conventional down-mix to a matrixed surround compatible down-mix is facilitated by a MTX conversion block applied as *post-processing* stage of the encoding tree.

The MTX conversion block has two inputs and two outputs. The two output signals are linear combinations of the two input signals. The resulting 2×2 processing matrix is dynamically varying and depends on the spatial parameters resulting from the spatial encoding process. If the surround channels contain relatively little energy, the two output signals of the MTX processing stage are (almost) identical to the two input signals. If, on the other hand, there is a significant surround activity, the 2×2 matrix creates negative crosstalk to signal surround activity to a matrixed surround decoder. The advantage of employing this process on a stereo down mix rather than on the multi-channel input,

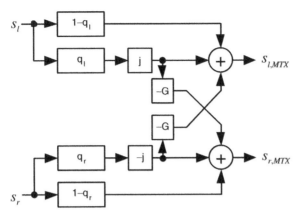

Figure 6.7 Matrixed surround conversion block. Reproduced by permission of the Audio Engineering Society, Inc, New York, USA.

is that the 2×2 processing matrix is *invertible*. In other words, the MPEG Surround decoder can 'undo' the processing by employing the inverse of the encoder matrix. As a result, the matrixed surround compatibility has no negative effect on the 5.1-channel reconstruction of an MPEG Surround decoder.

The matrixed surround conversion block is outlined in Figure 6.7. Both the down-mix signals, s_l and s_r, are split in two parts using parameters q_L and q_R. These parameters represent the relative amount of surround energy in each parameter band of s_l and s_r, respectively and are derived from the encoded spatial parameters. For nonzero q, part of the input signal is processed by a 90° phase shifter (indicated by the 'j' block). The phase-shifted signal is subsequently mixed out-of-phase to both output channels $s_{l_{MTX}}, s_{r_{MTX}}$, including a (fixed) weight $G = \frac{1}{\sqrt{3}}$ for the cross-term.

The scheme depicted in Figure 6.7 can be described in matrix notation employing a conversion matrix $\mathbf{V}_b$:

$$\begin{bmatrix} s_{l_{MTX}}, m(k) \\ s_{r_{MTX}}, m(k) \end{bmatrix} = \mathbf{V}_b \begin{bmatrix} s_{l,m}(k) \\ s_{r,m}(k) \end{bmatrix} = \begin{bmatrix} v_{11,b} & v_{12,b} \\ v_{21,b} & v_{22,b} \end{bmatrix} \begin{bmatrix} s_{l,m}(k) \\ s_{r,m}(k) \end{bmatrix} \qquad (6.16)$$

with

$$v_{11,b} = \frac{1 - q_{l,b} + j q_{l,b}}{\sqrt{1 - 2q_{l,b} + 2q_{l,b}^2}} \qquad (6.17)$$

$$v_{12,b} = \frac{j q_{r,b}}{\sqrt{3(1 - 2q_{r,b} + 2q_{r,b}^2)}} \qquad (6.18)$$

$$v_{21,b} = \frac{-j q_{l,b}}{\sqrt{3(1 - 2q_{l,b} + 2q_{l,b}^2)}} \qquad (6.19)$$

$$v_{22,b} = \frac{1 - q_{r,b} - j q_{r,b}}{\sqrt{1 - 2q_{r,b} + 2q_{r,b}^2}} \qquad (6.20)$$

The elements of the conversion matrix $\mathbf{V}$ are determined for each parameter band and are dependent on the spatial parameters resulting from the preceding tree:

$$q_{l,b} = f\left(\gamma_{1,b}\right)\left[\frac{\sqrt{P_{2,b}}}{\sqrt{P_{1,b}} + \sqrt{P_{2,b}}}\right]_{TTO_0} \tag{6.21}$$

$$q_{r,b} = f\left(\gamma_{2,b}\right)\left[\frac{\sqrt{P_{2,b}}}{\sqrt{P_{1,b}} + \sqrt{P_{2,b}}}\right]_{TTO_1} \tag{6.22}$$

$$f(\gamma) = \begin{cases} 1 & \text{if } |\gamma| > 1 \\ -1 - 2\gamma & \text{if } -1 > \gamma > -0.5 \\ \frac{1+2\gamma}{3} & \text{otherwise} \end{cases} \tag{6.23}$$

where X_{TTO_i} denotes the fraction X of TTO block E_i of Figure 6.6.

External down-mix analysis block

In some cases, the use of an externally provided down-mix may be preferred over an automated down-mix. For example, a studio engineer might produce separate stereo and multi-channel mixes from the same (multi-track) recording. MPEG Surround provides the possibility to transmit such an externally provided down-mix instead of the automated down-mix. In order to minimize potential differences in the resulting multi-channel reconstruction, the external down-mix analysis block parameterizes the differences between the automated and externally provided down-mixes. The external down-mix analysis block is used as a post-processor of the full spatial encoder tree. For each internal, automated down-mix channel $s_{i,m}$ and the corresponding externally provided down-mix channel $e_{i,m}$, the energy ratio within each parameter band is extracted according to:

$$\Delta L_{s_i e_i, b} = 10 \log_{10}\left(\frac{p_{s_{i,m}}}{p_{e_{i,m}}}\right) \tag{6.24}$$

This down-mix gain parameter describes the level adjustment in each parameter band that should be applied to the externally provided down-mix to result in a down-mix that is equal to the automated down-mix from a (statistical) energy point of view. On top of this ICLD parameter, residual signals $e_{i,m}$ can be transmitted for a user-selectable bandwidth to obtain waveform reconstruction of the automated down-mix from the (transmitted) external down-mix:

$$d_{i,m}(k) = s_{i,m}(k) - \eta \sqrt{\frac{p_{s_{i,m}}}{p_{e_{i,m}}}} e_{i,m}(k) \tag{6.25}$$

The parameter η controls the method of coding of the residual signal; $\eta = 0$ results in absolute coding of the automated down mix $s_{i,m}$, while for $\eta = 1$, the difference between the automated down-mix $s_{i,m}$ and the gain-adjusted externally provided down-mix $e_{i,m}$ is used as residual signal. The latter method is especially beneficial if there exists a high correlation between the externally provided down mix and the automated down-mix.

6.3.5 Parameter quantization and coding

Parameter quantization

For ICLD and ICC parameters, the same quantizer is used as applied in parametric stereo coders. The CPC coefficients are quantized linearly with a step size of 0.1 and a range between -2.0 and $+3.0$.

Further bitrate reduction techniques

The quantizer described in Section 6.3.5 aims at just-inaudible differences in spatial properties. An additional quantization strategy is also supplied based on a reduced number of quantizer steps to reduce the entropy per transmitted spatial parameter. This 'coarse' quantization comprises only every even quantizer index of the quantizer described in Section 6.3.5.

If such coarse quantization steps are applied, there is a risk that the relatively large discrete steps in changes in spatial properties give rise to audible artifacts. For example, if a certain sound object in the multi-channel content is slowly moving from one speaker location to another, the smooth movement in the original content may be reproduced at the decoder side as a sequence of discrete positions, each perceived position corresponding to a quantizer value. To resolve such artifacts, the encoder may signal a 'smoothing flag' in the bitstream, which signals the decoder to apply a low-pass filter on the discrete parameter values to result in a smooth transition between different quantizer values.

A related technique for further bitrate reduction is referred to as 'energy-dependent quantization'. This method allows for combinations of fine and coarse parameter quantization, depending on the amount of signal energy within the tree structure. If the amount of signal energy in a certain part of the parameter tree is significantly lower than the overall signal energy, large quantization errors in that specific part are in most cases inaudible, since they will be masked by signal components from other channels. In such cases, a very coarse parameter quantization can be applied for relatively weak channel pairs, while a fine quantization may be applied for strong (loud) channel pairs.

Besides changes in quantizer granularity, MPEG Surround also features the possibility to transmit only a selected number of parameters. More specifically, only a single ICC parameter may be transmitted instead of a separate ICC value for each TTO block. If this single ICC mode is enabled, the same transmitted ICC value is used in each OTT decoding block.

Finally, the resulting quantizer indexes are differentially encoded over time and frequency. Entropy coding is employed on the differential quantizer indexes to exploit further redundancies.

6.3.6 Coding of residual signals

As described in Section 6.3.4, TTO and TTT encoding blocks can generate residual signals. These residual signals can be encoded in a bit-efficient manner and transmitted along with the corresponding down-mix and spatial parameters.

Residual data do not necessarily need to be transmitted since MPEG Surround decoders are capable of reconstructing decorrelated signals with similar properties to those of the residual signals without requiring any additional information (see Section 6.4.2). However, if full waveform reconstruction at the decoder side is desired, residual signals can be transmitted. The bandwidth of the residual signals can be set at the encoder side, so that a trade-off can be made between bitrate consumption and reconstruction quality. The residual signals are transformed from the hybrid QMF to an MDCT representation and subsequently encoded into an AAC bitstream element. The residual part can be stripped from existing bitstreams to allow for bitstream scalability (without the need for re-encoding).

6.4 MPEG Surround decoder

6.4.1 Structure

The MPEG Surround decoder structure is outlined in Figure 6.8. The down-mix is first processed by a pre-gain, which is the inverse of the post-gain of the MPEG Surround encoder. Subsequently, the input signals are processed by an analysis filterbank that is identical to the filterbank described in Section 6.3.3. A spatial decoder regenerates multi-channel audio by reinstating the spatial properties described by the decoded parameters. Finally, applying a set of synthesis filterbanks and post-gains (the inverse of the encoder pre-gains) results in the time domain multi-channel output signals.

6.4.2 Spatial decoder

Operation principle

The spatial decoder generates multi-channel output signals from the down-mixed input signal by reinstating the spatial cues captured by the spatial parameters. The spatial synthesis of OTT decoding blocks employs so-called *decorrelators* and matrix operations

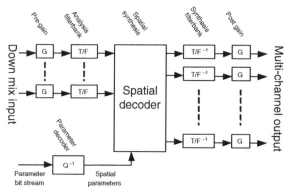

Figure 6.8   Structure of the MPEG Surround decoder. Reproduced by permission of the Audio Engineering Society, Inc, New York, USA.

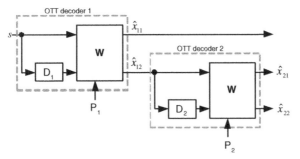

Figure 6.9 Concatenation of two OTT decoding blocks to achieve three-channel output. Reproduced by permission of the Audio Engineering Society, Inc, New York, USA.

in a similar fashion as parametric stereo decoders [47]. In a OTT decoding block, two output signals with the correct spatial cues are generated by mixing a mono input signal with the output of a decorrelator that is fed with that mono input signal.

Given the tree structures that were explained in Section 6.3.4, a first attempt for building a multi-channel decoder could be to simply concatenate OTT decoding blocks according to the tree structure at hand. An example of such a concatenation of OTT decoding blocks for three-channel output is shown in Figure 6.9. A mono input signal s is processed by a first decorrelator D_1 and an up-mix matrix $\mathbf{W}(P_1)$ to obtain two output signals $\hat{x}_{11}, \hat{x}_{12}$ with spatial parameters P_1:

$$\begin{bmatrix} \hat{x}_{11}(k) \\ \hat{x}_{12}(k) \end{bmatrix} = \mathbf{W}(P_1) \begin{bmatrix} s(k) \\ D_1(s(k)) \end{bmatrix} \tag{6.26}$$

with

$$\mathbf{W}(P_i) = \begin{bmatrix} w_{11}(P_i) & w_{12}(P_i) \\ w_{21}(P_i) & w_{22}(P_i) \end{bmatrix} \tag{6.27}$$

Signal $\hat{x}_{12}$ is subsequently processed by a second decorrelator D_2, and mixed with $\hat{x}_{12}$ itself based on a second spatial parameter set P_2 to generate two output signals $\hat{x}_{21}, \hat{x}_{22}$:

$$\begin{bmatrix} \hat{x}_{21}(k) \\ \hat{x}_{22}(k) \end{bmatrix} = \mathbf{W}(P_2) \begin{bmatrix} \hat{x}_{12}(k) \\ D_2(\hat{x}_{12}(k)) \end{bmatrix} \tag{6.28}$$

The up-mix matrices $\mathbf{W}$ ensure that their output pairs have the correct level difference as well as the correct correlation.

The correct output levels are obtained by scaling the two output signals according to the transmitted ICLD parameter, while the correct coherence is obtained by mixing the two input signals under the assumption that the output of each decorrelator gives an output that is statistically independent from its input, while having the same temporal and spectral signal envelopes. Consequently, both outputs $\hat{x}_{11}, \hat{x}_{12}$ of up-mix matrix $\mathbf{W}(P_1)$ will in many cases comprise signal portions of both inputs (i.e., the input signal s and the output of decorrelator D_1). Output signal $\hat{x}_{12}$ is subsequently processed by a second decorrelator D_2 as input for up-mix matrix $\mathbf{W}(P_2)$.

This scheme has the important drawback of decorrelators connected in series: the output of decorrelator D_1 is (partly) fed into decorrelator D_2. Given the most important requirement of decorrelators to generate output that is statistically independent from its input, its processing will result in a delay and temporal and/or spectral smearing of the input signals. In other words, the spectral and temporal envelopes of an input signal may be altered considerably, especially if the decorrelator contains reverberation-like structures. If two decorrelators are connected in series, the degradation of signal envelopes will be substantial. Moreover, since spatial parameters are temporally varying, temporal smearing and delays will cause an asynchrony between the signals and their parameters. This asynchrony will become larger if decorrelators are connected in series. Thus, concatenation of decorrelators should preferably be avoided.

Fortunately, the problem of concatenated decorrelators can be solved without consequences for spatial synthesis. Decorrelator D_2 should generate a signal that is statistically independent from $\hat{x}_{12}$, which is a combination of s and the output of decorrelator D_1. In other words, the output of D_2 should be independent of both s and the output of decorrelator D_1. This can be achieved by feeding decorrelator D_2 with mono input signal s instead of $\hat{x}_{12}$, if the decorrelators D_1 and D_2 are *mutually independent*. This enhancement is outlined in Figure 6.10.

The input of decorrelator D_2 is now obtained directly from s with a gain $\lambda_2(P_1)$ which compensates for the change in energy that would otherwise be caused by matrix $\mathbf{W}(P_1)$:

$$\lambda_i^2(P_1) = w_{i1}^2(P_1) + w_{i2}^2(P_1) \tag{6.29}$$

Furthermore, it can be observed that signal $\hat{x}_{12}$, which is a linear combination of s and the output of decorrelator D_1, is processed by matrix $\mathbf{W}(P_2)$ without any intermediate decorrelation process. Given the linear properties of the two matrix operations, the contribution of s within $\hat{x}_{21}$ and $\hat{x}_{22}$ can be obtained by a single (combined) matrix operation by multiplication of the respective elements from $\mathbf{W}(P_1)$ and $\mathbf{W}(P_2)$. The statistical equivalence of both schemes can be shown by computing the covariance matrices of the output signals in both cases, which are identical. In summary, cascaded decorrelators can be shifted through preceding OTT decoding blocks without changing statistical properties such as signal levels and mutual correlations, under the assumption that the different decorrelators are mutually independent.

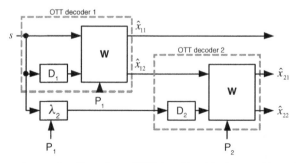

Figure 6.10 Enhanced concatenation of two OTT decoding blocks to achieve three-channel output with decorrelators in parallel. Reproduced by permission of the Audio Engineering Society, Inc, New York, USA.

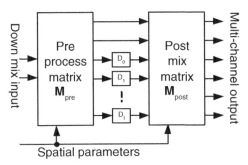

Figure 6.11 Generic spatial decoder. Reproduced by permission of the Audio Engineering Society, Inc, New York, USA.

The process of transforming spatial parameterization trees from cascaded decorrelator structures to decorrelators in parallel, extended with combined matrix multiplications leads to the *generalized* spatial decoder structure as shown in Figure 6.11. Any encoder tree configuration can be mapped to this generalized decoder structure. The input signals are first processed by a pre-process matrix $\mathbf{M}_{pre}$, which applies decorrelator input gains as outlined in Figure 6.10, TTT-type decoding (in case of a stereo down-mix), as well as any decoder-side *inversion* processes that should be applied on the down-mix (see Section 6.2.2). The outputs of the pre-matrix are fed to a decorrelation stage with one or more mutually independent decorrelators. Finally a post-mix matrix $\mathbf{M}_{post}$ generates the multi-channel output signals. In this scheme, both the pre-process matrix as well as the post-mix matrix are dependent on the transmitted spatial parameters.

Decorrelators

In all tree configurations some outputs of the mix-matrix $\mathbf{M}_{pre}$ are fed into decorrelators. These decorrelators create an output that is uncorrelated with their input. Moreover, in the case multiple decorrelators are used, they are conditioned such that their outputs will also be mutually uncorrelated (see Section 6.4.2). Figure 6.12 shows a diagram of the decorrelator processing that is performed on the hybrid domain signals.

The decorrelators comprise a delay (that varies in different frequency bands), a lattice all-pass filter, and an energy adjustment stage. The configuration for the delay and all-pass

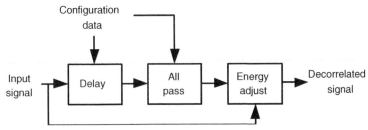

Figure 6.12 Diagram of decorrelator processing on hybrid QMF domain signals. Reproduced by permission of the Audio Engineering Society, Inc, New York, USA.

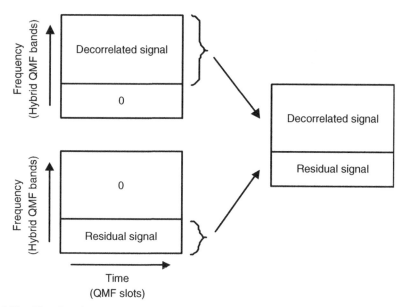

Figure 6.13 The decoder can generate a decorrelated signal with similar properties to those of the residual signal in frequency bands for which no residual is transmitted.

filter are controlled by the encoder using decorrelator configuration data. The all-pass coefficients of the different decorrelators were selected such that their output are mutually independent (even if the same signal is used as input).

In order to avoid audible reverberation in the case of transients, an energy adjustment stage scales the output of the decorrelator to match the energy level of the input signal in all frequency (processing) bands.

If residual signals are transmitted for certain OTT or TTT decoding blocks, the outputs of the corresponding decorrelators are replaced by the decoded residual signals. This replacement is only applied for the frequency range of the transmitted residual signal. For the remaining bandwidth, the decorrelator output is maintained. This process is visualized in Figure 6.13.

OTT decoding block

The up-mix matrix **W** for an OTT decoding block is determined by the following constraints:

1. The correlation of the two output signals must obey the transmitted ICC parameter.

2. The power ratio of the two output signals must obey the transmitted ICLD parameter.

3. The sum of the energies of the output signals must be equal to the energy of the input signal. Note that this constraint is slightly different from parametric stereo, where the *mean* of the energies of the output signals is equal to the energy of the input signal.

Given these three constraints, the 2×2 matrix $\mathbf{W}$ has one degree of freedom. One interpretation of this degree of freedom is a common rotation angle of the two output signals in a two-dimensional space spanned by the two input signals, in a similar way to that outlined in Section 5.5.2. The mix matrix $\mathbf{W}$ can be expressed using a common rotation angle β, a differential rotation angle α and two vector lengths λ_1 and λ_2:

$$\begin{bmatrix} \hat{x}_{1,m}(k) \\ \hat{x}_{2,m}(k) \end{bmatrix} = \begin{bmatrix} \lambda_1 \cos(\alpha + \beta) & \lambda_1 \sin(\alpha + \beta) \\ \lambda_2 \cos(-\alpha + \beta) & \lambda_2 \sin(-\alpha + \beta) \end{bmatrix} \begin{bmatrix} s_{1,m}(k) \\ D(s_{1,m}(k)) \end{bmatrix} \tag{6.30}$$

Minimization of $w_{12} + w_{22}$ leads to the following solution for β:

$$\beta = \tan\left(\frac{\lambda_2 - \lambda_1}{\lambda_2 + \lambda_1} \arctan(\alpha)\right) \tag{6.31}$$

with α depending on the ICC parameter as outlined in Section 5.5.2, Equation 5.25. The variables λ_1 and λ_2, representing the relative amplitudes of the two output signals with respect to the input, are given by:

$$\lambda_1 = \sqrt{\frac{10^{(\Delta L_{x_1 x_2}/10)}}{1 + 10^{(\Delta L_{x_1 x_2}/10)}}} \tag{6.32}$$

$$\lambda_2 = \sqrt{\frac{1}{1 + 10^{(\Delta L_{x_1 x_2}/10)}}} \tag{6.33}$$

The solution for β implies that $w_{12,i} = -w_{22,i}$. In other words, the synthesis matrix can also be written for each parameter band b as:

$$\mathbf{W}_b = \begin{bmatrix} \lambda_{1,b} \cos(\alpha_b + \beta_b) & +1 \\ \lambda_{2,b} \cos(-\alpha_b + \beta_b) & -1 \end{bmatrix} \begin{bmatrix} 1 & 0 \\ 0 & \lambda_{1,b} \sin(\alpha_b + \beta_b) \end{bmatrix} \tag{6.34}$$

Stated differently, the decorrelation signal level is identical in both output signals but the contribution to both output channels is in anti phase. Hence, this decoder synthesis matrix employs the same decomposition that was used at the encoder side (see Section 6.3.4), with the exception that the common out-of-phase component is now synthetically generated by decorrelation and scaling (with $\lambda_1 \sin(\alpha + \beta)$).

OTT decoding block using residual coding

If for a certain parameter band a residual signal $d_{1,m}$ is transmitted, the decorrelator output is *replaced* by the transmitted residual signal and the corresponding matrix elements are set to $+1$ and -1, respectively, according to the corresponding signal decomposition at the encoder (see Section 6.3.4):

$$\mathbf{W}_b = \begin{bmatrix} \lambda_{1,b} \cos(\alpha_b + \beta_b) & +1 \\ \lambda_{2,b} \cos(-\alpha_b + \beta_b) & -1 \end{bmatrix} \tag{6.35}$$

The fixed matrix weights for the residual signal of $+1$ and -1 (instead of the dynamic, parameter-dependent weights for the decorrelator outputs) make the system more robust against signal-reconstruction errors due to interpolation of mixing matrices (see Section 6.4.2). At the same time, the replacement strategy of residual signal and decorrelator output make the system scalable on a bitstream level. Due to the fact that the matrix elements for the down-mix remain the same (with or without residual signal), stripping of the residual signal from an encoded parameter stream results in a decoder output that is identical to the output that would have been obtained if encoding was performed without residual signals.

TTT decoding block using prediction mode

Three output signals $\hat{x}_l$, $\hat{x}_r$, $\hat{x}_c$ are synthesized according to the inverse encoder-side down-mix matrix using an estimated signal $\hat{s}_c$:

$$\begin{bmatrix} \hat{x}_{l,m}(k) \\ \hat{x}_{r,m}(k) \\ \hat{x}_{c,m}(k) \end{bmatrix} = \frac{1}{3} \begin{bmatrix} 2 & -1 & 1 \\ -1 & 2 & 1 \\ \sqrt{2} & \sqrt{2} & -\sqrt{2} \end{bmatrix} \begin{bmatrix} s_{l,m}(k) \\ s_{r,m}(k) \\ \hat{s}_{c,m}(k) \end{bmatrix} \tag{6.36}$$

with

$$\hat{s}_{c,m}(k) = \gamma_{1,b} x_{l,m}(k) + \gamma_{2,b} x_{r,m}(k) + d_{1,m}(k) \tag{6.37}$$

m the filter band index, b the processing band index and $d_{1,m}$ the residual signal. The resulting up-mix matrix $\mathbf{W}$ is then given by:

$$\begin{bmatrix} \hat{x}_{l,m}(k) \\ \hat{x}_{r,m}(k) \\ \hat{x}_{c,m}(k) \end{bmatrix} = \frac{1}{3} \begin{bmatrix} \gamma_{1,b} + 2 & \gamma_{2,b} - 1 & 1 \\ \gamma_{1,b} - 1 & \gamma_{2,b} + 2 & 1 \\ \sqrt{2}(1 - \gamma_{1,b}) & \sqrt{2}(1 - \gamma_{2,b}) & -\sqrt{2} \end{bmatrix} \begin{bmatrix} s_{l,m}(k) \\ s_{r,m}(k) \\ d_{1,m}(k) \end{bmatrix} \tag{6.38}$$

If no residual signal was transmitted, the resulting energy loss can be compensated for in two ways, depending on the complexity of the decoder. The first, low-complexity solution is to apply a gain to the three output signals according to the prediction loss. In that case, the up-mix matrix is given by:

$$\mathbf{W}_b = \frac{1}{3c_b} \begin{bmatrix} \gamma_{1,b} + 2 & \gamma_{2,b} - 1 & 0 \\ \gamma_{1,b} - 1 & \gamma_{2,b} + 2 & 0 \\ \sqrt{2}(1 - \gamma_{1,b}) & \sqrt{2}(1 - \gamma_{2,b}) & 0 \end{bmatrix} \tag{6.39}$$

This method does ensure correct overall power, but the relative powers of the three output signals, as well as their mutual correlations, may be different from those of the original input signals.

Alternatively, the prediction loss can be compensated for by means of a decorrelator signal. In that case, the (synthetic) residual signal $d_{1,m}$ of Equation (6.38) is generated by decorrelators fed by the two down-mix signals (only for those frequency bands for which no transmitted residual signal is available). This more complex method reconstructs the full covariance structure of the three output signals.

TTT decoding block based on energy reconstruction

TTT decoding based on energy reconstruction (henceforth called energy mode) supports two methods. These methods are characterized by the way the up-mix matrix is derived, using the same (transmitted) parameters. The bitstream header signals which method should be used.

In the *energy mode without center subtraction*, the left and right output signal are calculated from the left and right down-mix signal, respectively. In other words, the left output signal is generated independently from the right input channel and vice versa. The center signal is a linear combination of both down-mix signals. This method should be used if at least a certain frequency range the legacy stereo coder does not have waveform-preserving properties (for example when using SBR). The up-mix process is given by:

$$
\begin{bmatrix} \hat{x}_{l,m}(k) \\ \hat{x}_{r,m}(k) \\ \hat{x}_{c,m}(k) \end{bmatrix} = \begin{bmatrix} w_{11,b} & 0 \\ 0 & w_{22,b} \\ w_{31,b} & w_{32,b} \end{bmatrix} \cdot \begin{bmatrix} s_{l,m}(k) \\ s_{r,m}(k) \end{bmatrix}
$$

(6.40)

The matrix elements are given by:

$$
w_{11,b} = \sqrt{\frac{\kappa_{1,b} \cdot \kappa_{2,b}}{\kappa_{1,b} \cdot \kappa_{2,b} + \kappa_{2,b} + 1}}
$$

(6.41)

$$
w_{22,b} = \sqrt{\frac{\kappa_{1,b}}{\kappa_{1,b} + \kappa_{2,b} + 1}}
$$

(6.42)

$$
w_{31,b} = \frac{1}{2} \cdot \sqrt{2 \cdot \frac{\kappa_{2,b} + 1}{\kappa_{1,b} \cdot \kappa_{2,b} + \kappa_{2,b} + 1}}
$$

(6.43)

$$
w_{32,b} = \frac{1}{2} \cdot \sqrt{2 \cdot \frac{\kappa_{2,b} + 1}{\kappa_{1,b} + \kappa_{2,b} + 1}}
$$

(6.44)

with

$$
\kappa_{i,b} = 10^{\Delta L_{i,b}/10}.
$$

(6.45)

The *energy mode with center subtraction*, on the other hand, tries to improve the reconstruction of the left and right signals by utilizing cross-terms. This method is especially beneficial if the core coder is at least partly preserving the waveforms of its input. More details on this method are given in [126].

MTX inversion block

If the transmitted down-mix is encoded using a matrixed surround conversion block (see Section 6.3.4), the stereo input signal is processed by a matrixed surround inversion matrix **W** which is the *inverse* of the encoder-side conversion matrix **V**:

$$\mathbf{W}_b = \mathbf{V}_b^{-1} \tag{6.46}$$

External down-mix inversion block

If an external down mix was provided, the external down-mix inverter aims at reconstructing the (discarded) automated down-mix from the transmitted external down-mix. The reconstructed down-mix signal $\hat{s}_{i,m}$ for channel i is given by:

$$\hat{s}_{i,m}(k) = \begin{bmatrix} \eta\sqrt{\kappa_{i,b}} & 1 \end{bmatrix} \begin{bmatrix} e_{i,m}(k) \\ d_{i,m}(k) \end{bmatrix} \tag{6.47}$$

with $\kappa_{i,b}$ dependent on the external down-mix gain parameter $\Delta L_{s_i e_i, b}$ according to Equation (6.45) for parameter band b and down-mix channel i, $e_{i,m}$ the transmitted external down mix, $d_{i,m}$ the external down-mix residual for channel i (if available) and η is computed using the decision regarding absolute or relative coding of the residual signals (if available).

Matrix elements for a mono down-mix

The construction of pre- and post-mix matrices for the mono-based tree as outlined in the left panel of Figure 6.5 is outlined in Figure 6.14. The gain compensation factors for decorrelator inputs resulting from cascaded OTT blocks are applied in the pre-mix matrix $\mathbf{M}_{pre}$. The LFE signal is not subject to decorrelation and hence its output signal is solely constructed using gain factors resulting from all respective OTT blocks. If an external down-mix was provided, the external down-mix inversion block is combined with $\mathbf{M}_{pre}$ as well (not shown in Figure 6.14).

The mixing matrices **W** for each OTT decoding block are combined in a single post-mix matrix $\mathbf{M}_{post}$. This process can be performed for any OTT tree structure, including trees with more than six input or output channels.

Matrix elements for a stereo down-mix

The construction of the pre- and post-mix matrices for a stereo-based tree is shown in Figure 6.15. The pre-mix matrix comprises the combined effect of matrixed surround inversion (MTX) or external-down mix inversion (EXT) and the TTT decoding process. The left and right outputs of the TTT output signals are subsequently fed to parallel decorrelators. The post-mix matrix is then composed of three parallel OTT blocks. The OTT decoding block for the center and the LFE channel does not have a decorrelator

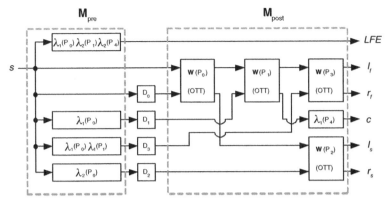

Figure 6.14 Pre- and post-matrix construction for a mono-based tree configuration. Reproduced by permission of the Audio Engineering Society, Inc, New York, USA.

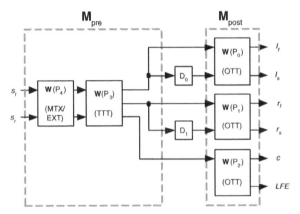

Figure 6.15 Pre- and post-matrix construction for a stereo-based tree configuration. Reproduced by permission of the Audio Engineering Society, Inc, New York, USA.

input since no correlation synthesis between center and LFE is applied (i.e., the respective ICC values are set to +1).

Parameter positions and interpolation

For each transmitted parameter set the mixing matrices are determined as described previously. Similar to the approach pursued in Parametric Stereo (see Chapter 5), these matrices correspond in most cases to a single time instance, which depends on the segmentation and windowing procedure of the encoder. For QMF sample indices (k) in between parameter positions, the mixing matrices $\mathbf{M}_{pre}$ and $\mathbf{M}_{post}$ are interpolated linearly (i.e. its real and imaginary parts are interpolated individually). This interpolation of mixing matrices has the advantage that the decoder can process each 'slot' of hybrid QMF samples (i.e. one sample from each sub-band) one by one, without the need of storing a whole frame of sub-band samples in memory. This results in a significant memory reduction compared to frame-based synthesis methods.

6.4.3 Enhanced matrix mode

MPEG Surround features an analysis element that is capable of estimating spatial parameters based on a conventional or matrixed surround compatible down-mix. This element
enables MPEG Surround to work in a mode that is similar to matrixed surround systems,
i.e., by means of a matrixed surround compatible down-mix without transmission of additional parameters, or alternatively, to generate multi-channel representations from legacy
stereo material. For such a mode, the MPEG Surround decoder analyzes the transmitted
(stereo) down-mix and generates spatial parameters that are fed to the spatial decoder to
up-mix the stereo input to multi-channel output. Alternatively, this analysis stage can be
employed already in the *encoder* to enable multi-channel audio transmission in MPEG
Surround format based on conventional stereo source material.

A spatial decoder using this enhanced matrix mode is shown in Figure 6.16. The spatial
parameters required to compute the matrix elements of the pre- and post-mix matrix are
generated by an analysis module A. The analysis module measures two parameters of the
received down-mix for each parameter band. These parameters are the down-mix level
difference $\Delta L_{s_1 s_2, b}$ and the down-mix cross-correlation $c_{s_1 s_2, b}$. To avoid analysis delays,
these parameters are estimated using first-order filtering involving data from the past.

Three (parameter-band dependent) states are updated whenever a new slot of down-
mix signals is received. These states comprise the signal power $p_{s_i, b}$ of each down-mix
channel:

$$p_{s_i, b}(k) = \varepsilon p_{s_i, b}(k - 1) + (1 - \varepsilon) \sum_{m=m_b}^{m_{b+1}-1} s_{i,m}(k) s_{i,m}^*(k) \qquad (6.48)$$

as well as the cross-spectrum $\chi_{s_1 s_2, b}^2(k)$:

$$\chi_{s_1 s_2, b}^2(k) = \varepsilon \chi_{s_1 s_2, b}^2(k - 1) + (1 - \varepsilon) \mathrm{Re} \left(\sum_{m=m_b}^{m_{b+1}-1} s_{1,m}[k] s_{2,m}^*(k) \right) \qquad (6.49)$$

The coefficient ε determines the adaptation speed, which is based on a time constant
of $T = 60$ ms:

$$\varepsilon = \exp \left(\frac{-64}{T f_s} \right) \qquad (6.50)$$

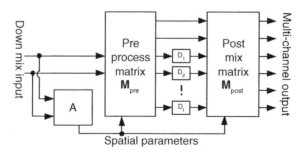

Figure 6.16 MPEG Surround spatial decoder using the enhanced matrix mode. Reproduced by
permission of the Audio Engineering Society, Inc, New York, USA.

Every fourth slot, the current states $p_{s_i,b}$ and $\chi^2_{s_1s_2,b}$ are converted to a level difference $\Delta L_{s_1s_2,b}$ and a normalized cross-correlation coefficient $\rho_{s_1s_2,b}$:

$$\Delta L_{s_1s_2,b}(k) = 10\log_{10}\frac{p_{s_1,b}(k)}{p_{s_2,b}(k)} \tag{6.51}$$

$$\rho_{s_1s_2,b}(k) = \frac{\chi_{s_1s_2,b}(k)}{\sqrt{p_{s_1,b}(k)p_{s_2,b}(k)}} \tag{6.52}$$

These down-mix parameters are subsequently converted to indices for a table lookup of the spatial parameters (i.e., for the matrixed surround inversion stage, and the subsequent TTT and OTT decoding elements) required for multi-channel reconstruction. The parameter *position* for this newly generated parameter set equals the current analysis position plus four slots (i.e., $k + 4$).

The lookup tables resulted from a 'training process' in which a very large set of multi-channel material was encoded using an MPEG Surround encoder using the MTX conversion block. Subsequently, the spatial parameters that resulted from the encoding process, as well as the *down-mix parameters* $\Delta L_{s_1s_2,b}$ and $\rho_{s_1s_2,b}$ were analysed and stored in a database. From this database, histograms were constructed for the distribution for each MPEG Surround parameter given a specific combination of the down-mix parameters of a certain time–frequency tile. Such a histogram (normalized to obtain an estimate of the probability distribution function, or PDF) is shown in Figure 6.17. All values for the CLD and ICC between the left front and left surround channels were selected in case the down-mix CLD ($\Delta L_{s_1s_2,b}$) was between -0.5 and $+0.5$ dB, and the down-mix correlation ($\rho_{s_1s_2,b}$) was between -0.05 and $+0.05$. The observations were pooled across frequency bands and analysis frames and across all encoded audio excerpts. The resulting PDF for the front-surround CLD is shown in the top panel of Figure 6.17, while the PDF for the front-surround ICC is shown in the lower panel.

Interestingly, both the CLD and ICC between l_f and l_s have a distribution with a clearly identifiable peak value. In other words, given $\Delta L_{s_1s_2,b} = 0$ and $\rho_{s_1s_2,b} = 0$, certain spatial parameter values between l_f and l_s seem to occur more often than others. Thus, if spatial parameters are unknown, a best guess for the CLD and ICC between l_f and l_s in this case would comprise a certain metric to describe the most probable value given the PDFs shown in Figure 6.17, such as their mean or mode. This best guess can then be obtained for a wide range of down-mix parameters, and for each MPEG Surround parameter individually. The result of such a procedure is outlined in Figure 6.18. The CLD between l_f and l_s (represented by different grey shades) is given as a function of the down-mix parameters. Only positive values for the down-mix CLD are shown.

Figure 6.18 has several interesting properties. For example, if the down-mix has a correlation close to -1, and both down-mix channels have approximately equal power (represented by the top left corner of the figure), the front-surround CLD is in most cases between -15 and -25 dB. In other words, the surround channels contained considerable more energy than the front channels. This is in line with what one would expect from a matrixed surround encoder (and thus the MTX conversion stage): the surround channels are mixed in anti-phase in the down-mix, hence resulting in a negative correlation if the surround channels are dominant in terms of energy.

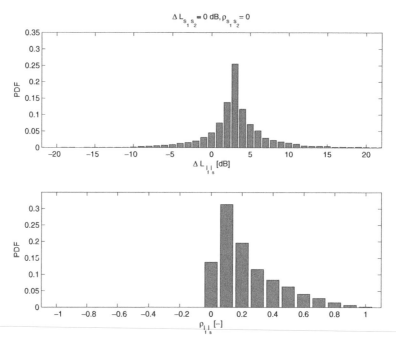

Figure 6.17 Probability distribution function for the CLD (top panel) and ICC (lower panel) between l_f and l_s in case the down-mix parameters amount to $\Delta L_{s_1 s_2,b} = 0$ and $\rho_{s_1 s_2,b} = 0$.

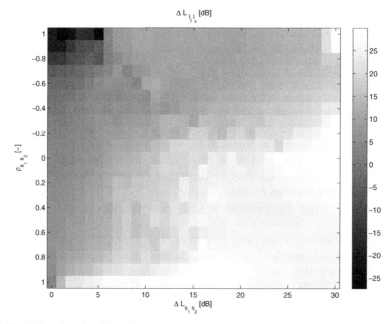

Figure 6.18 Value for the CLD between l_f and l_s as a function of the down-mix parameters $\Delta L_{s_1 s_2,b}$ and $\rho_{s_1 s_2,b}$.

On the other hand, if the down-mix correlation is close to +1 and there exists a strong level difference between the down-mix channels (lower-right corner), the CLD between front and surround channels is positive, indicating that the surround channels are virtually inactive. For down-mix parameters in between, there exists a gradual change between front-only and surround-only activity, with an exception of the top-right corner, which indicates front-only activity.

6.5 Subjective evaluation

During the MPEG Surround development, the progress and corresponding performance have been documented in detail in several publications [49, 121, 122, 265] and documented in a formal verification test report [144]. The results published in those papers primarily focused on bitrate scalability, different channel configurations, support for external down mixes, and binaural decoding (see also Chapter 8).

The purpose of the listening tests described in this chapter is to demonstrate that existing stereo services can be upgraded to high-quality multi-channel audio in a fully backward compatible fashion at transmission bit rates that are currently used for stereo. In a first test, the MPEG Surround performance is demonstrated using two different core coders (AAC and MP3), and a comparison is made against alternative systems to upgrade a stereo transmission chain to multi-channel audio. In a second test, the performance for the operation mode without transmission of spatial parameters (i.e. using the enhanced matrix mode) is outlined.

6.5.1 Test 1: operation using spatial parameters

Stimuli and method

The list of codecs that were employed in the test is given in Table 6.2. The total employed bit rate (160 kbps) was set to a value that is commonly used for high-quality stereo transmission.

Configuration (1) represents stereo AAC at 128 kbps in combination with 32 kbps of MPEG Surround (MPS) parametric data. Configuration (2) is based on a different core coder (MP3 in combination with MPEG Surround) using a slightly lower parametric

Table 6.2 Codecs under test.

Configuration	Codec	Core bitrate (kbps)	Spatial bitrate (kbps)	Total bitrate (kbps)
1	AAC stereo + MPS	128	32	160
2	MP3 + MPS	149	11	160
3	MP3 Surround	144	16	160
4	AAC stereo + DPLII	160	n/a	160
5	AAC Multichannel	160	n/a	160

bitrate (and consequently a slightly higher bitrate for the core coder; informal listening indicated that this resulted in a higher overall quality). Configuration (3) is termed 'MP3 Surround' [120] which is a proprietary extension to the MPEG-1 layer 3 (MP3) codec. This extension also employs parametric side information to retrieve multi-channel audio from a stereo down-mix, but is not compatible with MPEG Surround. Configuration (4) employs the Dolby Prologic II matrixed surround system (DPLII) for encoding and decoding in combination with stereo AAC at a bit rate of 160 kbps. Configuration (5) is AAC in multi-channel mode, which represents state-of-the-art discrete channel coding.

For configurations (1), (4) and (5), state-of-the-art AAC encoders were used. For configurations (2) and (3), an encoder and decoder available from www.mp3surround.com have been used (version April 2006). Dolby Prologic II encoding and decoding was performed using the Dolby-certified 'Minnetonka Surcode for Dolby Prologic II' package (version 2.0.3) using its default settings.

Eight listeners participated in this experiment. All listeners had significant experience in evaluating audio codecs and were specifically instructed to evaluate the overall quality, consisting of the spatial audio quality as well as any other noticeable artifacts. In a double-blind MUSHRA test [148], the listeners had to rate the perceived quality of several processed excerpts against the original (i.e. unprocessed) excerpts on a 100-point scale with 5 anchors, labeled 'bad', 'poor', 'fair', 'good' and 'excellent'. A hidden reference and a low-pass filtered anchor (cut-off frequency at 3.5 kHz) were also included in the test. The subjects could listen to each excerpt as often as they liked and could switch in real time between all versions of each excerpt. The experiment was controlled from a PC with an RME Digi 96/24 sound card using ADAT digital out. Digital-to-analog conversion was provided by an RME ADI-8 DS 8-channel D-to-A converter. Discrete pre-amplifiers (Array Obsydian A-1) and power amplifiers (Array Quartz M-1) were used to feed a 5.1 loudspeaker setup employing B&W Nautilus 800 speakers in a dedicated listening room according to ITU recommendations [147].

A total of 11 critical excerpts were used as listed in Table 6.3. The excerpts are the same as used in the MPEG Call for Proposals (CfP) on Spatial Audio Coding [142], and range from pathological signals (designed to be critical for the technology at hand) to movie sound and multi-channel productions. All input and output excerpts were sampled at 44.1 kHz.

Results

The subjective results of each codec and excerpt are shown in Figure 6.19. The horizontal axis denotes the excerpt under test, the vertical axis the mean MUSHRA score averaged across listeners, and different symbols indicate different codecs. The error bars denote the 95% confidence intervals of the means.

For all excerpts, the hidden reference (square symbols) has scores virtually equal to 100 with a very small confidence interval. The low-pass anchor (circles), on the other hand, consistently has the lowest scores around 10–20. The scores for AAC multi-channel (rightward triangles) are between 20 and 60 for the individual excerpts, and its average rates approximately 40. Stereo AAC in combination with Dolby Prologic II (leftware triangles) scores only slightly higher on average. For 10 out of the 11 excerpts, the combination of stereo AAC and MPEG Surround has the highest scores (diamonds).

Table 6.3 Test excerpts.

Excerpt	Name	Category
1	BBC applause	Pathological/ambience
2	ARL applause	Pathological/ambience
3	Chostakovitch	Music
4	Fountain music	Pathological/ambience
5	Glock	Pathological
6	Indie2	Movie sound
7	Jackson1	Music
8	Pops	Music
9	Poulenc	Music
10	Rock concert	Music
11	Stomp	Music (with LFE)

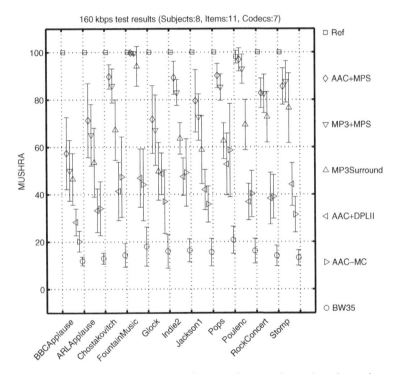

Figure 6.19 Mean subjective results averaged across listeners for each codec and excerpt. Error bars denote 95% confidence intervals. Reproduced by permission of the Audio Engineering Society, Inc, New York, USA.

The overall scores (averaged across subjects and excerpts) are given in Figure 6.20. AAC with MPEG Surround scores approximately 5 points higher than MP3 with MPEG Surround. MP3 Surround scores approximately 15 points lower than MPEG Surround when combined with MP3.

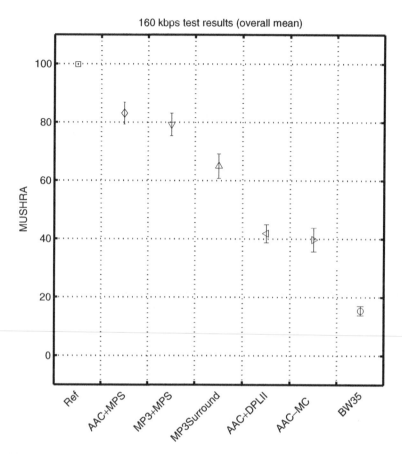

Figure 6.20 Overall mean subjective results for each codec. Reproduced by permission of the Audio Engineering Society, Inc, New York, USA.

Discussion

The results indicate the added value of parametric side information with a stereo transmission channel (configurations (1), (2) and (3) vs configurations (4) and (5)). The increase in quality for MPEG Surround compared with discrete multi-channel coding or matrixed surround methods amounts to more than 40 MUSHRA points (using AAC as core coder), which is a considerable improvement. All three parameter-enhanced codecs demonstrated such a clear benefit, enabling high-quality audio transmission at bitrates that are currently used for high-quality stereo transmission. The two core coders tested seem to have only a limited effect, since the difference between AAC with MPEG Surround and MP3 with MPEG Surround is reasonably small. On the other hand, given the large difference in quality between configurations (4) and (5) which are based on the same core coder using virtually the same bitrate, the two different parametric enhancements (MPEG Surround and MP3 Surround, respectively) seem to differ significantly in terms of quality *and* compression efficiency; MPEG Surround delivers significantly higher quality while using only 69% of the parameter bit rate of MP3 Surround.

6.5.2 Test 2: operation using enhanced matrix mode

Stimuli and method

The list of configurations that were employed in the test is given in Table 6.4.

All configurations employed stereo AAC at a bitrate of 160 kbps as core coder. Configuration (1) serves as a high-quality anchor employing MPEG Surround with 32 kbps of spatial parameter data. For configuration (2), no parameters were transmitted; the MPEG Surround encoder generated a matrixed surround compatible stereo down-mix (using the MTX conversion block), while the MPEG Surround decoder operated in enhanced matrix mode (EMM) as outlined in Section 6.4.3. Configuration (3) employed a Dolby Prologic II (DPLII) encoder to convert multi-channel to matrixed surround compatible stereo, and a Dolby Prologic II decoder to resynthesize multi-channel signals. The same AAC and Prologic encoders and decoders were employed as in the previous test. The test procedure and reproduction setup were also equal to those in the previous test. The 10 excerpts that were used are given in Table 6.5. These items were used in the MPEG Surround verification test of which the results are given in [144].

Results

The MUSHRA scores averaged across excerpts and subjects are given in Figure 6.21. The square symbols denote the reference, which has as score of 100. Configuration (1),

Table 6.4 Codecs under test.

Configuration	Codec	Core bitrate (kbps)	Spatial bitrate (kbps)	Total bitrate (kbps)
1	AAC stereo + MPS	160	32	192
2	AAC stereo + MPS EMM	160	n/a	160
3	DPLII	160	n/a	160

Table 6.5 Test excerpts.

Excerpt	Name	Category
1	Bonobo	Movies drama
2	Elliot	Movies drama
3	Lavilliers	Pop music (with LFE)
4	Ravel	Orchestra
5	SantaCruz	Pop music
6	StationAtmo	Ambience
7	Tennis	Ambience
8	Thalheim	Pop music
9	Tower	Jazz music
10	Violin	Orchestra

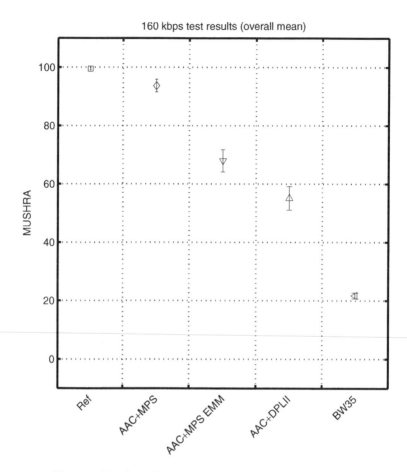

Figure 6.21   Overall mean subjective results for each codec.

employing MPEG Surround using 32 kbps of parametric overhead, results in a very high subjective quality with a score of over 90 (diamonds). If no parametric data has been transmitted, as in configuration (2) denoted by the downward triangles, the average score drops to around 70. Dolby Prologic II (upward triangles), on the other hand, has a score of around 55. The low-pass filtered anchor (leftward triangles) has a score around 23.

Discussion

The results indicate a similar benefit of transmitting spatial parameters as the previous test. The additional transmission of 32kbps of parametric data results in a significant increase in perceptual quality compared with the two systems under test that do not employ any transmission of any additional information.

Configurations (2) and (3) both employ a matrixed surround compatible down-mix to facilitate multi-channel reconstruction at the decoder side. Interestingly, despite the very similar method to convey and signal surround activity of these two configurations, the MPEG Surround system is capable of reconstructing a multi-channel signal that is significantly closer to the original signal than the Dolby Prologic II system. Hence even if the transmission of additional data is undesirable or even impossible, MPEG Surround achieves a very competitive multi-channel experience.

6.6 Conclusions

A parametric extension to mono or stereo audio codecs has been described that provides high-quality multi-channel capabilities at bitrates that are equal to those currently employed for stereo transmission. The subjective listening test revealed superior perceptual quality of MPEG Surround over conventional multi-channel AAC, matrixed surround, and MP3 Surround coders at an overall bit rate of 160 kbps.

Full backward compatibility is guaranteed with legacy receivers by storing parametric side information in the ancillary data part of existing compression schemes. The spatial side information is scalable between 0 and (typically) 32 kbps, although higher rates are supported for applications demanding (near) transparency.

The system described is very flexible in terms of the input and output channels; all common speaker configurations are supported. The flexibility also extends to the down-mix domain. MPEG Surround features automated down mixes that can be mono, stereo, or matrixed surround compatible stereo. Even multi-channel 5.1 can be used as a down-mix for configurations with a higher number of audio channels (e.g., 7.1 or 10.2). Furthermore, support is provided for externally provided down-mixes as well. Last but not least, MPEG Surround features an enhanced matrix mode that enables up-conversion of legacy stereo material to high-quality multi-channel content.

7

Binaural Cues for a Single Sound Source

7.1 Introduction

Convolution using head-related transfer functions (HRTFs) is a widely used method to evoke the percept of a virtual sound source at a given spatial position. As described in Chapter 3, the use of HRTFs requires a database of pairs of impulse responses, preferably matched to the anthropometric properties of the user. Because of the large amount of data present in individual HRTF sets that is normally required to generate externalized virtual sound sources, it is desirable to find an efficient representation of HRTFs. For example, attempts have been made to only measure HRTF sets for a limited range of source positions and to interpolate HRTFs for positions in between (based on the magnitude transfers [270], spherical harmonics [81], eigentransfer functions [56], pole-zero approximations [27] or spherical spline methods [54]). Other studies described HRTFs by deriving a small set of basis spectra with individual, position-dependent weights [56, 59, 162]. Although these methods are sound in physical terms, there is a risk that the basis functions that are very important in terms of the least-squares error of the fit are not so relevant in terms of human auditory perception. Another, more psycho-acoustically motivated approach consisted of determining the role of spectral and inter-aural phase cues present in the HRTFs. Wightman and Kistler [274] showed that low-frequency inter-aural time differences dominate in sound localization, while if the low frequencies are removed from the stimuli, the apparent direction is determined primarily by inter-aural level differences and pinna cues. Hartmann and Wittenberg [115] and Kulkarni et al. [170, 171] showed that the frequency-dependent ITD of anechoic HRTFs can be simplified by a frequency-independent delay without perceptual consequences. Huopaniemi and Zacharov [131] discussed three methods to reduce HRTF information. The first method entailed smoothing of the HRTF magnitude spectra by a rectangular smoothing filter with a bandwidth equal to the equivalent rectangular bandwidth (ERB) [97]. Similar experiments using gammatone transfer functions were performed by Breebaart and Kohlrausch [42, 43]. The second method embodied weighting of the errors in an HRTF approximation with the inverse of the ERB scale as weighting function. The third method used frequency warping to account for the

Spatial Audio Processing: MPEG Surround and Other Applications   Jeroen Breebaart and Christof Faller
© 2007 John Wiley & Sons, Ltd

nonuniform frequency resolution of the auditory system. From many of these studies, it can be concluded that, although a frequency-independent ILD does not result in an externalized image, the complex magnitude and phase spectra which are present in HRTFs can be simplified to some extent without deteriorating the externalization.

The approach that is pursued here is to exploit limitations of the binaural hearing system to reduce the amount of information to describe HRTFs. It is assumed that the spatial audio coding approach that was so far applied to stereo and multi-channel audio signals can be applied to HRTF impulse responses as well. More specifically, it is hypothesized that the inter-aural and spectral (envelope) properties of anechoic HRTFs can be 'downsampled' to an ERB-scale resolution without perceptual consequences. Furthermore, it is hypothesized that for anechoic HRTFs, the absolute phase spectrum is irrelevant and only the relative (inter-aural) phase between the two impulse responses has to be taken into account. These assumptions lead to a very simple parametric description of HRTFs, that comprises an average IPD or ITD per frequency band and two signal level parameters that describe the average signal level in each band for each of the two ears. This set can be extended with an IC parameter for each frequency band; however for many *anechoic* HRTFs the IC parameter is often very close to $+1$ and hence small deviations from $+1$ can be ignored for the anechoic case.

In Section 7.2, the HRTF parameterization procedure will be described in more detail. Subsequently, three different listening tests will be described to evaluate the HRTF parameterization process and to assess the number of parameters required for a perceptually transparent HRTF representation.

7.2 HRTF parameterization

The parameterization method that is outlined below employs several novel concepts. First, instead of *processing* existing HRTFs phase or magnitude spectra, a parameter-based analysis and synthesis approach is pursued using perceptually relevant transformation. Second, the proposed method enables modification of the amount of parameters to represent HRTFs in an adaptive fashion. Third, it is based on *inter-aural* phase relationships only, while completely discarding the *absolute* phase characteristic of HRTFs. Fourth, a comparison can be made between spectrally smooth and step-wise approximations.

7.2.1 HRTF analysis

The HRTF analysis step extracts parameters from an head-related impulse response (HRIR) pair of a specific spatial position. In a first step, the HRIR impulse responses $h_l(n)$, $h_r(n)$ are converted to frequency domain HRTFs using an M-point FFT, resulting in frequency-domain HRTFs $H_l(m)$, $H_r(m)$. Subsequently, the parameters are extracted that characterize the HRTFs by a set of perceptually motivated basis functions. The basis functions form a set of bandpass filters that mimic the known (spectral) limitations of the human auditory system. Given their bandpass characteristic, the basis functions are referred to as *parameter bands*. Each parameter band (or basis function) has an associated parameter band index b ($b = 0, \ldots, B - 1$). The parameter band basis functions are specified in a matrix $\mathbf{Q}$ that has M rows and B columns. Each column

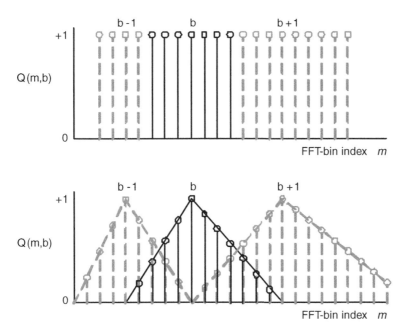

Figure 7.1 Parameter band basis functions for nonoverlapping bands (top panel) and overlapping bands (lower panel). Only nonzero values are shown for clarity.

(i.e. each parameter band index) specifies the parameter band filter characteristic (i.e. the form of the basis function) along FFT bin (frequency) index m. For example, a certain parameter band b may comprise a certain limited range of FFT bin indexes $m = m_b, \ldots, m_{b+1} - 1$. In that case, all values of **Q** in row b are zero, except for the columns $m = m_b, \ldots, m_{b+1} - 1$, which are set to nonzero values (for example $+1$). Two examples of parameter band basis functions are given in Figure 7.1. The entries $Q(m, b)$ of the matrix **Q** are given for parameter bands $b - 1$, b and $b + 1$ (only nonzero values are shown for clarity). The horizontal axes denotes the FFT-bin index m; the vertical axis represents the value $Q(m, b)$. The top panel shows $Q(m, b)$ for *nonoverlapping* parameter bands, i.e. each FFT index is associated with one unique parameter band b. In this case, due to the absence of any (spectral) overlap of parameter bands, the parameter band basis functions are *orthogonal*. The lower panel demonstrates *overlapping* parameter bands. In that case, the values $Q(m, b)$ comprise discretely sampled triangular shapes that are centered on the center frequency (or center FFT-bin) of a parameter band. The sum of all values $Q(m, b)$ across parameter bands b equals $+1$. Furthermore, it can be observed that the parameter bandwidth is different for the three parameter bands $(b - 1, b, b + 1)$.

The center frequencies $f_c(b)$ (in Hz) of each parameter band b are spaced according to a warped ERB scale:

$$f_c(b) = 228.7 \left(10^{0.00437wb} - 1\right) \tag{7.1}$$

with w the warp factor. For $w = +1$, the parameter bandwidths and center frequencies f_c of each parameter band b exactly follow the ERB scale [98]. For larger values of w, the center frequencies are spaced wider and consequently, the parameter bandwidths become wider with a factor w, and the number of parameter band basis functions B *decreases*.

For a given warp factor w and a given FFT length M, the ERB-scale frequencies f_c for each parameter band b are converted to FFT-bin indices m_b (using the nearest integer index, while at the same time ensuring a minimum 'width' of 1 FFT bin for each band b).

In the next step, the various parameter band basis functions are used to model the spectral envelope of each HRTF, as well as the inter-aural phase characteristic. For example, the power $p_{h_i,b}$ of HRTF H_i can be extracted for each parameter band b:

$$p_{h_i,b} = \sum_m Q^+(m,b) H_i(m) H_i^*(m) \tag{7.2}$$

where matrix Q^+ is the pseudo-inverse of the parameter band matrix:

$$Q^+ = \left(Q^T Q\right)^{-1} Q^T. \tag{7.3}$$

In a similar fashion, the IPD $\phi_{h_l h_r,b}$ for parameter band b can be extracted using:

$$\phi_{h_l h_r,b} = \angle\left(\sum_m Q^+(m,b) H_l(m) H_r^*(m)\right) \tag{7.4}$$

A comparison between original HRTF spectra and extracted parameters is shown in Fig 7.2. The top panels represent the magnitude spectra of the left ear (top left panel) and right ear (top right panel); the lower panel represents the interaural phase angle. The solid lines are the magnitude spectra from the original HRTF (subject '3' from the CIPIC database [2], for an elevation of 0 and azimuth of 65° to the left), the circles represent the parameter values and are given as a function of the center frequency of the respective basis function. The parameters were extracted using overlapping parameter bands and $w = 2.0$ (resulting in $B = 20$ bands). As can be observed from the top panels, the parameter frequencies are approximately linearly spaced on a logarithmic axis. Furthermore, the match between parameter values and magnitude spectra is quite accurate for low frequencies, while at high frequencies, some fine-structure details in the magnitude spectra are not represented by the coarsely sampled parameter values.

7.2.2 HRTF synthesis

The reconstructed complex-valued HRTF spectra $\hat{H}_l, \hat{H}_r$ are obtained by reinstating the extracted parameters ($p_{h_l,b}, p_{h_r,b}$, and $\phi_{h_l h_r,b}$) on (the spectrum of) a Dirac impulse using the basis functions Q. If during analysis stage, overlapping bands were employed, the same overlapping bands were used during HRTF reconstruction.

In principle, IC parameters should be extracted as well, but given the high IC values for the HRTF set under test, the IC parameters were assumed to be sufficiently close to +1 and hence no dedicated HRTF decorrelation procedure was required. It should be

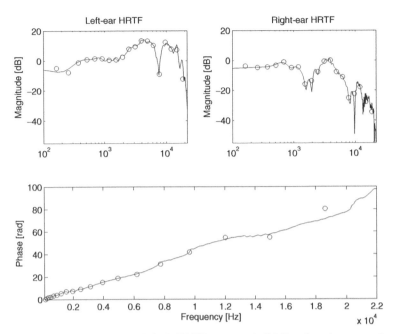

Figure 7.2 Comparison between original HRTF spectra (solid lines) and extracted parameters (circles) for the left-ear magnitude spectrum (top left panel), the right-ear magnitude spectrum (top right panel) and the interaural phase (bottom panel).

noted however that this can not be generalized to arbitrary HRTF sets. More detailed information on HRTF parameter extraction and synthesis can be found in [202].

Two examples of reconstructed magnitude spectra are shown in Figure 7.3. The original (dotted line) and the reconstructed (solid line) HRTF magnitude spectra for the left ear are shown using overlapping bands (top panel) or nonoverlapping bands (lower panel). Comparison of the two panels in Figure 7.3 clearly reveals the step-wise approach of the nonoverlapping bands, while the overlapping bands give a smooth (interpolated) magnitude spectrum.

7.3 Sound source position dependencies

7.3.1 Experimental procedure

In a first experiment, the aim was to determine the minimum number of parameter band basis functions required to represent the left and right HRTFs for a specified position to obtain a fully transparent representation of an HRTF pair.

As outlined in Section 7.2, Equation (7.1), the warp factor w directly influences the number of parameter bands used for describing the parameterized HRTF spectra. With an increase of w, the number of parameter bands decreases and hence the error or difference

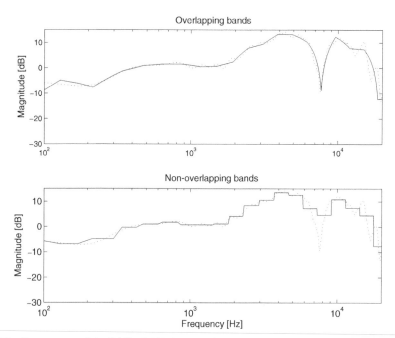

Figure 7.3   Reconstructed (solid line) HRTF magnitude spectra (left-ear only) using overlapping parameter bands (top panel) or nonoverlapping bands (lower panel). The dotted lines indicate the original HRTF magnitude spectra.

Table 7.1   Elevation and azimuth (in degrees) for the nine positions under test.

	1	2	3	4	5	6	7	8	9
Elevation Θ	0	90	−22.5	11.25	135	191.25	157.5	−33.75	33.75
Azimuth Φ	0	0	−10	−55	−20	10	45	65	25

between original HRTF spectra and parameterized HRTF spectra will increase. In this subjective evaluation, the parameterized HRTF pair was compared with the original HRTF pair for the same location. During the experiment, the value of w was changed adaptively to determine a threshold value. This experiment was performed for both overlapping as well as nonoverlapping parameter bands.

A set of nine different spatial positions were selected. These nine selected sound source locations are listed in Table 7.1. The *horizontal polar coordinate system* was employed to indicate azimuth Φ and elevation Θ. A publicly available HRTF database was used (CIPIC, [2]).

For each subject, thresholds were determined for each of the nine positions and for both parameterization methods (overlapping and nonoverlapping). The stimulus was a 300 ms burst of pink noise, followed by 200 ms of silence. Within each trial, the same noise realization was used for each of the three intervals. A new noise realization was generated for each trial.

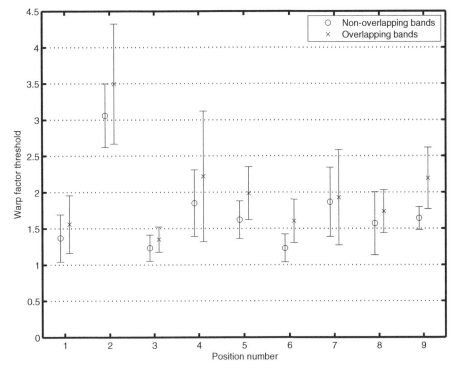

Figure 7.4 Thresholds (y-axis) of the warp factor w as a function of the nine sound source locations (x-axis). Different symbols indicate different parameterization methods (see text). Error bars denote the standard deviation of the results across subjects and trials.

7.3.2 Results and discussion

The results of the listening test are shown in Figure 7.4. The thresholds of the warp factor w are shown as a function of the sound source position, with the numbers 1 to 9 referring to the positions given in Table 7.1. The results are averaged across subjects and repetitions; the error bars denote the standard deviations of the results across all respective trials. The circles indicate results for the nonoverlapping parameterization method, while the crosses represent results for overlapping (triangular) parameter bands.

As can be observed from Figure 7.4, position number 2 gives a remarkably higher threshold value (between 3 and 3.5) than the other positions, which vary between 1.2 and 2.2. Both parameterization methods show a very similar trend, although the threshold values for the overlapping parameter method are up to about 0.5 ERB higher than for the nonoverlapping method. Interestingly, none of the average threshold values is below 1.0 ERB.

The averaged threshold that was observed in the experiments has a value that is larger than $+1$, for all positions and both methods. Moreover, values below $+1$ were only observed for the nonoverlapping parameterization method. This result strongly supports our hypothesis that a 1-ERB spectral resolution is sufficient to represent HRTFs without any perceptual degradation.

The threshold for the warp factor w also depends on the position of the sound source. For position number 2 ($\Theta = 90°$, $\Phi = 0°$), the average threshold was found to be higher than for all other positions in the test. The smallest w values were found for positions 1 ($\Theta = 0°$, $\Phi = 0°$), 3 ($\Theta = -22.5°$, $\Phi = -10°$) and 6 ($\Theta = 191.25°$, $\phi = 10°$). These positions are all close to the median plane. A possible explanation for this result is that the spectral shape of the HRTFs for these positions seems to have more pronounced details, resulting in a higher number of parameter bands to fully describe all perceptually relevant attributes. For position 2, on the other hand, the HRTF spectra are relatively smooth and can hence be modeled with a relatively limited set of parameters (resulting in a high threshold value).

Finally, there exists a systematic difference between the two parameterization methods. The triangular, overlapping parameter bands consistently result in higher thresholds (2.02 on average) than the orthogonal, rectangular parameter bands ($w = 1.69$, i.e., the latter requires more parameter bands for a perceptually identical representation). The reason for this finding might be related to the step-wise approximation of the HRTF spectrum for the nonoverlapping method compared to a much more smooth HRTF approximation for the overlapping parameter bands. Possibly the actual error between approximated and actual HRTF spectrum is larger for the nonoverlapping parameterization method, or subjects could use the sharp spectral edges as cues to detect stimuli resulting from the nonoverlapping parameterization method more easily.

7.4 HRTF set dependencies

7.4.1 Experimental procedure

In this experiment, HRTFs measured for different subjects are compared to investigate potential effects of various HRTF sets. Four different HRTF sets of the CIPIC HRTF database were evaluated in this test. These are the sets '018', '137' and '147' as well as the set '003' that was already used in the previous experiment. The comparison of these HRTF sets was done for positions 1 and 9. Position 1 was selected because it resulted in a low w threshold in the previous experiment and it represents a very common sound source position. Position 9 was selected since it is positioned more to the side and has a slightly different elevation component.

The experimental procedure, the subjects and the stimuli were identical to the procedure employed in the previous test. Only one parameterization method (using overlapping parameter bands) was used to determine the thresholds for w.

7.4.2 Results and discussion

The results are shown in Figure 7.5. The two tested positions for each of the four HRTF sets are marked on the abscissa. The corresponding thresholds of the warp factor w averaged across three trials of the five subjects is given along the ordinate. The errorbars denote the standard deviation of all measurements.

The results for HRTF set '003' are quite similar to the data obtained from the previous experiment; position 1 has a threshold value of about 1.4, while position 9 results in a

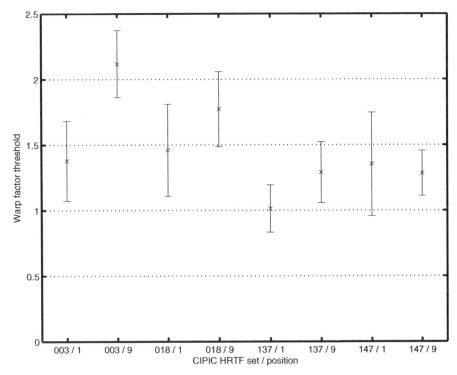

Figure 7.5   Thresholds for the warp factor w for the four CIPIC HRTF sets ('003', '018' '137' '147') and two positions ('1' and '9').

threshold of about 2.1. For the other three HRTF sets, however, thresholds are generally lower for corresponding positions. A minimum of 1.0 is obtained for HRTF set '137', position 1. The tendency for HRTF sets '003', '018' and '137' that position 9 gives rise to higher thresholds is not observed for HRTF set '147'.

The mean thresholds averaged across subjects found in this experiment range from 1.0 to 2.1, a range that is still in line with our hypothesis that thresholds are expected to be at least 1.0. In this respect, position 1 of HRTF set '137' showed to be most critical.

The results also indicate that different HRTF sets may give rise to different threshold values, and that directional dependencies may differ across HRTF sets. In other words, there does not seem to be one single position or sound source direction that can be characterized by a low number of parameter bands in a general sense.

7.5 Single ITD approximation

In the previous experiments, the phase difference between corresponding HRTF pairs was parameterized as a frequency-dependent property. From physical considerations, it is likely that the inter-aural phase characteristic closely mimics a linear phase (delay) behavior, with some small deviations (see Figure 7.2). In the current experiment, we investigate whether the linear-phase approximation is feasible in combination with parameterized

HRTF spectra. In other words, the frequency-dependent interaural phase parameters are reduced to a single inter-aural time delay (ITD) parameter for each sound source position.

7.5.1 Procedure

Given the dominance of low frequencies in ITD-based sound source localization, the ITD for a specific HRTF pair was estimated from the low-frequency part only (i.e. from 0 to 1.5 kHz). The ITD parameter was estimated by minimizing the L2 norm of the inter-aural phase error within the specified frequency range (see [202] for details). The resulting ITD was subsequently used to generate the HRTF phase spectra by computing the corresponding phase difference at the center frequencies of each parameter band. HRTF analysis and synthesis was performed using overlapping parameter bands at a fixed spectral resolution of 1 ERB ($w = 1$).

Since the objective of this test was to investigate potential audibility of a linear-phase approximation, no adaptive measurement procedure was employed. Instead, subjects were presented with three intervals. One of these intervals contained a noise burst processed with a parameterized, linear-phase HRTF, while the two other intervals were convolved with the original, unmodified HRTFs. Subjects had to indicate the odd one out in a series of 30 trials for each of the 9 spatial positions. As a control condition, the same experiment was employed with a frequency-dependent phase parameterization (denoted by 'nonlinear phase'). The various sets of 30 trials for 9 positions and the two different conditions were presented in random order.

7.5.2 Results and discussion

The average (pooled across subjects) percentage correct responses for the nine different sound source positions are shown in Figure 7.6. The circles represent the linear-phase parameterization and the crosses denote the control condition using frequency-dependent phase parameters. The error bars denote the standard deviation of the mean (across subjects). The dashed vertical line represents the expected percentage correct based on chance; the dotted lines represent the corresponding 95% confidence interval.

The results indicate that for most positions, the linear-phase approximation method (circles) resulted in response rates of about 30–40% correct. This also holds for the control condition (squares). Furthermore, for most positions the inter-subject variability (represented by the error bars) is relatively constant. The only exception is position 4. For this position, the linear-phase approximation resulted in statistically significant higher correct responses than what one would expected based on pure guessing. This result is not observed for the control condition (crosses).

These results indicate that for most positions, the frequency-dependent phase parameters can be reduced to a single ITD parameter without significant perceptual consequences. The one exception is position 4. For this position, subjects had a score that is statistically significantly different from what one would expect based on chance. This effect is not observed for the corresponding control condition, which indicates that subjects

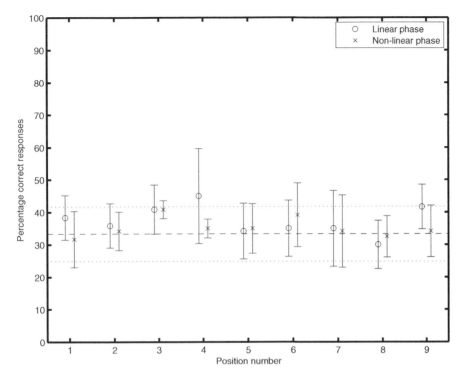

Figure 7.6 Average percentage of correct responses for nine different sound source positions. The circles correspond to the linear-phase approximation; the crosses to the frequency-dependent (nonlinear) phase method. Error bars denote the standard deviation across subjects. The dashed lines represent the 95% confidence intervals based on chance.

indeed used the different inter-aural phase cues for detecting the target. However, despite the statistical significance, the mean score for position 4 is still quite low (around 45% correct), which indicates that the cue that subjects used is quite weak.

7.6 Conclusions

The results presented in this chapter indicate that for non-individualized, anechoic HRTFs, significant data reduction can be achieved without perceptual consequences. The applied lossy HRTF representation method is based on known limitations of the human hearing system, and specifically exploits limitations in its frequency resolution.

It has been demonstrated that using noise stimuli, the absolute phase spectrum of the HRTF can be replaced by a linear phase curve, with an inter-aural delay that matches the average delay of the low-frequency part of an HRTF pair.

For the magnitude spectra, one magnitude value per ERB seems to be both a sufficient as well as a required resolution to result in a transparent HRTF representation. A coarser frequency resolution can result in audible differences for certain combinations of a specific HRTF pair and a subject. Although some HRTF pairs seem to be less sensitive to reductions in the number of parameters, comparisons between various HRTF sets and analysis of inter-subject results indicated that this can not be generalized.

In the next chapter, the proposed parametric method will be extended to multiple simultaneous sound sources, the incorporation of this method in MPEG Surround will be outlined, and listening test results using a variety of (multi-channel) audio content will be discussed.

8

Binaural Cues for Multiple Sound Sources

8.1 Introduction

The synthesis process of virtual auditory scenes has been an ongoing research topic for many years. The aim of so-called binaural rendering systems is to evoke the illusion of one or more sound sources positioned around the listener using stereo headphones. The positions of the sound sources can preferably be modified in terms of the perceived azimuth, elevation and distance. More advanced systems also include room acoustic models to simulate the acoustical properties, such as reflecting walls within the virtual space.

Binaural rendering has benefits in the field of research, simulation and entertainment [237]. Especially in the field of entertainment, the virtual auditory scene should sound very compelling and 'real'. In order to achieve such a realistic percept, several aspects have to be taken into account, such as the change in sound source positions with respect to head movement [197], room acoustic properties such as early reflections and late reverberation [236], and using system personalization to match the anthropometric properties of the individual user [198, 199, 276]. Because of the complex nature of current state-of-the-art systems, several concessions are required for feasible implementations (cf. [159]), especially if the number of sound sources that has to be rendered simultaneously is large.

Recent trends in consumer audio show a shift from stereo to multi-channel audio content, as well as a shift to solid state and mobile devices. These developments cause additional constraints on transmission and rendering systems. Firstly, the number of audio channels that has to be transmitted increases beyond two. The corresponding increase in transmission bandwidth for conventional, discrete-channel audio coders is often undesirable and sometimes even unavailable. Secondly, consumers often use headphones for audio rendering on a mobile device. To experience the benefit of multi-channel audio, a compelling binaural rendering system is required. This can be quite a challenge given the limited processing power and battery life of mobile devices.

In this chapter, a binaural rendering process will be described that exploits spatial audio coding techniques. The method is based on the analysis and synthesis of perceptually relevant parameters of a virtual auditory scene. The analysis and synthesis of these

Spatial Audio Processing: MPEG Surround and Other Applications   Jeroen Breebaart and Christof Faller
© 2007 John Wiley & Sons, Ltd

so-called binaural parameters is outlined in Sections 8.3 and 8.4; the integration of this method in the recently finalized MPEG Surround standard (cf. [138] and Chapter 6) for multi-channel audio compression is described in Section 8.5.

8.2 Binaural parameters

Binaural parameters refer to the effective spatial cues (i.e., ILD, ITD, ICC, and spectral envelopes) at the level of the eardrums that result from an auditory scene with one or more virtual sound sources. Binaural cues are the result of: (1) sound source spectral properties of each individual sound source; (2) the change in signal properties induced by HRTF convolution; and (3) the summation of signals from each sound source at the level of the ear drum. In Chapter 4, it was shown that spatial properties between audio channels can be described using spatial parameters, such as time- and frequency-dependent ICLD, ICTD/ICPD and ICC parameters. In Chapter 7, the same approach was outlined for binaural rendering, using ILD, IPD and ICC parameters between HRTF pairs.

The fact that inter-channel dependencies, as well as acoustical transfer characteristics can be described using a common repertoire of statistical (signal) properties allows for an efficient combination of spatial cecoding and binaural rendering in a single processing step, as will be outlined below. The underlying hypothesis is that spatial parameters, as well as HRTF parameters are locally stationary, which means that they are assumed to be constant within certain time/frequency tiles. In other words, it is assumed that HRTFs are parameterized according to the method described in Chapter 7.

8.3 Binaural parameter analysis

8.3.1 Binaural parameters for a single sound source

In conventional binaural rendering systems, a sound source i with associated time-domain signal $x_i(t)$ is rendered at a certain position by convolving the signal with a pair of head-related impulse responses $h_{l,i}(t)$, $h_{r,i}(t)$, for the left and right ears, respectively, to result in binaural signals $y_{l,i}(t)$, $y_{r,i}(t)$:

$$y_{v,i}(t) = \int_{\tau=0}^{\infty} x_i(t - \tau) h_{v,i}(\tau) \, d\tau, \tag{8.1}$$

with $v \in \{l, r\}$. This process is visualized in the left panel of Figure 8.1.

It is often convenient to express the convolution in the frequency domain using a frequency-domain representation $X_i(f)$ of a short segment of $x_i(t)$:

$$Y_{v,i}(f) = H_{v,i}(f) X_i(f) \tag{8.2}$$

with $H_{l,i}(f)$, $H_{r,i}(f)$ the frequency-domain representations (head-related transfer functions) of $h_{l,i}(t)$, $h_{r,i}(t)$, respectively. The power $p_{y_{v,i}}$ at the eardrum resulting from signal

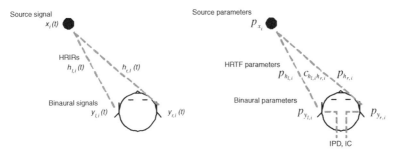

Figure 8.1 Synthesis of a virtual sound source by means of HRIR convolution (left panel) and by means of parametric representations (right panel). Reproduced from Breebaart, J. (2007). Analysis and synthesis of binaural parameters for efficient 3D audio rendering in MPEG Surround. IEEE Int. Conf. on Multimedia and Expo (ICME 2007), Beijing, China. Copyright 2007, IEEE.

$y_{v,i}$ in frequency band b is given by:

$$p_{yv,i,b} = \frac{1}{f(b+1) - f(b)} \int\limits_{f=f(b)}^{f(b+1)} Y_{v,i}(f)Y_{v,i}^*(f)\,\mathrm{d}f, \tag{8.3}$$

with $(^*)$ the complex conjugation operator, and $f(b)$ the lower edge frequency of frequency band b. For clarity and readability, the subscript b will not be given in the following equations; all described processing should nevertheless be performed in each parameter band individually. If the HRTF magnitude spectra $H_{v,i}(f)$ are locally stationary (i.e. constant within the frequency band b), this can be simplified to:

$$p_{yv,i} = p_{hv,i} p_{xi} \tag{8.4}$$

with $p_{hv,i}$ the power within frequency band b of HRTF $H_{v,i}$:

$$p_{hv,i} = \frac{1}{f(b+1) - f(b)} \int\limits_{f=f(b)}^{f(b+1)} H_{v,i}(f)H_{v,i}^*(f)\,\mathrm{d}f \tag{8.5}$$

and p_{xi} the power of the source signal $X_i(f)$ in frequency band b:

$$p_{xi} = \frac{1}{f(b+1) - f(b)} \int\limits_{f=f(b)}^{f(b+1)} X_i(f)X_i^*(f)\,\mathrm{d}f \tag{8.6}$$

Thus, given the local stationarity constraint, the power at the level of the eardrums follows from a simple multiplication of the power of the sound source and the power of the HRTF in corresponding frequency bands. In other words, statistical properties of binaural signals can be deducted from statistical properties of the source signal and from the HRTFs. This parameter-based approach is visualized in the right panel of Figure 8.1.

The inter-aural phase difference ϕ in parameter band b is given by the phase difference $\phi_{y_{l,i} y_{r,i}}$ between the signals $y_{l,i}$ and $y_{r,i}$ in parameter band b:

$$\phi = \phi_{y_{l,i} y_{r,i}} = \angle \left(\int_{f=f(b)}^{f(b+1)} Y_{l,i}(f) Y_{r,i}^*(f) \, df \right) \tag{8.7}$$

Under the assumption of local stationarity of inter-aural HRTF phase spectra, the IPD can be derived directly from the HRTF spectra themselves, without involvement of the sound source signal:

$$\phi = \phi_{h_{l,i} h_{r,i}} \tag{8.8}$$

with $\phi_{h_{l,i} h_{r,i}}$ the average phase angle of the HRTF pair corresponding to position i and frequency band b:

$$\phi_{h_{l,i} h_{r,i}} = \angle \left(\int_{f=f(b)}^{f(b+1)} H_{l,i}(f) H_{r,i}^*(f) \, df \right) \tag{8.9}$$

The equations above assume local stationarity of HRTF magnitude and inter-aural phase spectra to estimate the resulting binaural parameters. However, strong deviations from stationarity within analysis bands may result in a decrease in the inter-aural coherence (IC) for certain frequency bands, which can be perceived as a change in the spatial 'compactness' of a virtual sound source. To capture this property, the IC is estimated for each frequency band b. In the current context, the coherence is defined as the absolute value of the average normalized cross-spectrum:

$$c = c_{y_{l,i} y_{r,i}} = \frac{\left| \frac{1}{f(b+1)-f(b)} \int_{f=f(b)}^{f(b+1)} Y_{l,i}(f) Y_{r,i}^*(f) \, df \right|}{\sqrt{P_{y_{l,i}} P_{y_{r,i}}}} \tag{8.10}$$

The IC parameter has a dependency on the source signal x_i. For broadband signals, however, it's expected value however is only dependent on the HRTFs:

$$c = c_{h_{l,i} h_{r,i}} \tag{8.11}$$

with

$$c_{h_{l,i} h_{r,i}} = \frac{\left| \frac{1}{f(b+1)-f(b)} \int_{f=f(b)}^{f(b+1)} H_{l,i}(f) H_{r,i}^*(f) \, df \right|}{\sqrt{P_{h_{l,i}} P_{h_{r,i}}}} \tag{8.12}$$

In summary, under the local stationarity constraint, the binaural parameters p_{y_l}, p_{y_r}, IPD and IC resulting from a single sound source can be estimated from the sound source parameters p_{x_i} and the HRTF parameters $p_{h_{l,i}}$, $p_{h_{r,i}}$, $\phi_{h_{l,i}, h_{r,i}}$ and $c_{h_{l,i}, h_{r,i}}$.

8.3.2 Binaural parameters for multiple independent sound sources

For multiple simultaneous sound sources, conventional methods convolve each individual source signal i with an HRTF pair corresponding to the desired position, followed by summation:

$$Y_v(f) = \sum_i H_{v,i}(f)X_i(f) \tag{8.13}$$

Under the constraint of independent sound source signals $x_i(t)$, the power at the eardrums in frequency band b is given by the sum of the powers of each individual virtual sound source:

$$P_{y_v} = \sum_i P_{y_v,i} \tag{8.14}$$

which can be written for stationary HRTF properties as:

$$P_{y_v} = \sum_i P_{h_v,i} P_{x_i} \tag{8.15}$$

The net IPD ϕ resulting from the simultaneous virtual sound sources i is given by:

$$\phi = \phi_{y_{l,i} y_{r,i}} = \angle \left(\sum_i \int_{f=f(b)}^{f(b+1)} Y_{l,i}(f) Y_{r,i}^*(f) \, df \right) \tag{8.16}$$

This formulation can also be written in terms of parameters:

$$\phi = \angle \left(\sum_i e^{j\phi_{h_{l,i} h_{r,i}}} c_{h_{l,i} h_{r,i}} P_{x_i} \sqrt{P_{h_{l,i}} P_{h_{r,i}}} \right) \tag{8.17}$$

The IC can be estimated similarly:

$$c = \frac{\left| \sum_i e^{j\phi_{h_{l,i} h_{r,i}}} c_{h_{l,i} h_{r,i}} P_{x_i} \sqrt{P_{h_{l,i}} P_{h_{r,i}}} \right|}{\sqrt{P_{y_{l,i}} P_{y_{r,i}}}} \tag{8.18}$$

8.3.3 Binaural parameters for multiple sound sources with varying degrees of mutual correlation

The assumption of independent signals across various objects may hold for many applications, especially if each signal is associated with independent sound sources. However, for some applications, the various signals may comprise common components. For example if a virtual multi-channel audio setup is simulated, the signals that are radiated by the virtual loudspeakers may exhibit a significant mutual correlation. In that case, these correlations have to be taken into account in the binaural parameter estimation process. The

ICC for band b is denoted by $c_{x_{i_1}, x_{i_2}}$, for sound sources i_1 and i_2. In that case, the binaural parameters are estimated according to:

$$p_{y_v} = \sum_i \left(p_{h_{v,i}} p_{x_i} \right) + \dots \tag{8.19}$$

$$\sum_{i_1} \sum_{i_2 \neq i_1} \sqrt{r_{v i_1 i_2}} c_{x_{i_1}, x_{i_2}} \cos \left(\frac{\phi_{h_{l,i_1}} h_{r,i_1} - \phi_{h_{l,i_2}} h_{r,i_2}}{2} \right)$$

with

$$r_{v i_1 i_2} = p_{h_{v,i_1}} p_{h_{v,i_2}} p_{x_{i_1}} p_{x_{i_2}} c_{h_{l,i_1} h_{r,i_1}} c_{h_{l,i_2} h_{r,i_2}} \tag{8.20}$$

In a similar way, the IPD and IC are given by:

$$\phi_{y_l y_r} = \angle (\chi) \tag{8.21}$$

$$c_{y_l y_r} - \frac{|\chi_b|}{\sqrt{p_{y_l} p_{y_r}}} \tag{8.22}$$

with

$$\chi = \sum_i \left(e^{j \phi_{h_{l,i} h_{r,i}}} c_{h_{l,i} h_{r,i}} p_{x_i} \sqrt{p_{h_{l,i}} p_{h_{r,i}}} \right) + \dots \tag{8.23}$$

$$\sum_{i_1} \sum_{i_2 \neq i_1} \left(e^{\frac{j \phi_{h_{l,i_1}} h_{r,i_1} + j \phi_{h_{l,i_2}} h_{r,i_2}}{2}} c_{x_{i_1}, x_{i_2}} \sqrt{q_{i_1 i_2}} \right)$$

with

$$q_{i_1 i_2} = p_{h_{l,i_1}} p_{h_{r,i_2}} p_{x_{i_1}} p_{x_{i_2}} c_{h_{l,i_1} h_{r,i_1}} c_{h_{l,i_2} h_{r,i_2}} \tag{8.24}$$

In these equations, the IPD ϕ of each sound source is assumed to be distributed symmetrically across the two binaural signals (i.e. $\phi/2$ is the phase offset that is applied to the left-ear signal, and $-\phi/2$ is the phase offset of the right-ear signal). As can be observed, these equations are equivalent to those given in Section 8.3.2 for $c_{x_{i_1}, x_{i_2}} = 0$ if $i_1 \neq i_2$.

If the decrease in coherence due to HRTF convolution using different impulse responses for both ears is ignored (i.e. $c_{h_l h_r} = 1$) and hence it is assumed that the coherence of the binaural signal pair Y_L, Y_R is dominated by the fact that (partially) incoherent sources have different spatial positions, the estimation process simplifies to:

$$p_{y_v} = \dots \tag{8.25}$$

$$\sum_{i_1} \sum_{i_2} \sqrt{p_{h_v,i_1} p_{h_v,i_2} p_{x_{i_1}} p_{x_{i_2}}} c_{i_1,i_2} \cos \left(\frac{\phi_{h_{l,i_1}} h_{r,i_1} - \phi_{h_{l,i_1}} h_{r,i_1}}{2} \right)$$

$$\chi(b)(b) = \dots \tag{8.26}$$

$$\sum_{i_1} \sum_{i_2} \sqrt{p_{h_l,i_1} p_{h_r,i_2} p_{x_{i_1}} p_{x_{i_2}}} c_{i_1,i_2} e^{j \left(\frac{\phi_{h_{l,i_1}} h_{r,i_1} + \phi_{h_{l,i_1}} h_{r,i_1}}{2} \right)}$$

8.4 Binaural parameter synthesis

8.4.1 Mono down-mix

The synthesis process comprises reinstating the binaural parameters on a mono down-mix signal $x(t)$ of the object signals. Using a frequency-domain representation, one frame of the down-mix signal is given by:

$$X(f) = \sum_i X_i(f) \tag{8.27}$$

The power in each band of this down-mix signal frame, p_x, is then given by:

$$p_x = \sum_{i_1} \sum_{i_2} c_{x_{i_1} x_{i_2}} \sqrt{p_{x_{i_1}} p_{x_{i_2}}} \tag{8.28}$$

The reconstructed binaural signals $\hat{Y}_l$, $\hat{Y}_r$ are obtained using a matrix operation $\mathbf{W}_b$ that is derived for each parameter band (b) independently:

$$\begin{bmatrix} \hat{Y}_l(f) \\ \hat{Y}_r(f) \end{bmatrix} = \mathbf{W}_b \begin{bmatrix} X(f) \\ D(X(f)) \end{bmatrix} \tag{8.29}$$

with $D(.)$ a decorrelator which generates a signal that has virtually the same temporal and spectral envelopes as its input, but is independent of its input. The matrix coefficients ensure that for each frame, the two binaural output signals $\hat{Y}_l$, $\hat{Y}_r$ have the desired levels, IPD and IC relations. The solution for $\mathbf{W}_b$ is given by (see Chapter 5 for a detailed explanation of these equations):

$$\mathbf{W}_b = \begin{bmatrix} \lambda_l \cos(\alpha + \beta) & \lambda_l \sin(\alpha + \beta) \\ \lambda_r \cos(-\alpha + \beta) & \lambda_r \sin(-\alpha + \beta) \end{bmatrix} \tag{8.30}$$

with

$$\lambda_l = \sqrt{\frac{p_{yl}}{p_x}} e^{+j\phi_{yl\,yr}/2} \tag{8.31}$$

$$\lambda_r = \sqrt{\frac{p_{yr}}{p_x}} e^{-j\phi_{yl\,yr}/2} \tag{8.32}$$

$$\alpha = \frac{1}{2} \arccos\left(c_{yl\,yr}\right) \tag{8.33}$$

$$\beta = \tan\left(\frac{|\lambda_r| - |\lambda_l|}{|\lambda_r| + |\lambda_l|} \arctan(\alpha)\right) \tag{8.34}$$

8.4.2 Extension towards stereo down-mixes

In the previous sections, binaural parameters were analyzed and reinstated from a mono down-mix signal $X(f)$. For several applications, however, it is beneficial to provide means to extend the down-mix channel configuration to stereo. An example of a relevant application scenario is the synthesis of a virtual multi-channel 'home cinema setup' based on a conventional stereo down-mix signal, as will be outlined in more detail in Section 8.5. A popular down-mix matrix equation for down-mixing five-channel material to stereo is given by:

$$
\begin{bmatrix} Y_{l,ITU}(f) \\ Y_{r,UTU}(f) \end{bmatrix} = \begin{bmatrix} 1 & 0 & q & q & 0 \\ 0 & 1 & q & 0 & q \end{bmatrix} \begin{bmatrix} X_{lf}(f) \\ X_{rf}(f) \\ X_c(f) \\ X_{ls}(f) \\ X_{rs}(f) \end{bmatrix}
\tag{8.35}
$$

with $Y_{l,ITU}$, $Y_{r,ITU}$ the stereo down-mix signal pair, and X_{lf}, X_{rf}, X_c, X_{ls}, X_{rs} the signals for left-front, right-front, center, left-surround, right-surround, respectively, and $q = 1/\sqrt{2}$. Hence signals from loudspeakers positioned at the left side are only present in the left down-mix signal, and signals from loudspeakers positioned at the right side are only present in the right down-mix signal. The center channel is distributed equally over both down-mix channels.

The corresponding binaural signals Y_l, Y_r using HRTFs are given by:

$$
\begin{bmatrix} Y_l(f) \\ Y_r(f) \end{bmatrix} = \begin{bmatrix} H_{L,lf}(f) & H_{L,rf}(f) & H_{L,c}(f) & H_{L,ls}(f) & H_{L,rs}(f) \\ H_{R,lf}(f) & H_{R,rf}(f) & H_{R,c}(f) & H_{R,ls}(f) & H_{R,rs}(f) \end{bmatrix} \begin{bmatrix} X_{lf}(f) \\ X_{rf}(f) \\ X_c(f) \\ X_{ls}(f) \\ X_{rs}(f) \end{bmatrix}
\tag{8.36}
$$

When comparing Equations (8.35 and 8.36), two important differences can be observed. Firstly, the ITU down mix is frequency independent, while the HRTF matrix comprises frequency-dependent transfer functions. Secondly, the zero-valued cross terms in Equation (8.35) are replaced by nonzero HRTFs. If one assumes that the coherence of the virtual center channel (which is equal to the coherence between HRTFs $H_{l,c}$ and $H_{r,c}$) to be (sufficiently close to) $+1$, it can be shown that the coherence of the binaural signal pair Y_l, Y_r resulting from Equation (8.36) will be equal or *higher* than the coherence of the ITU-downmix signal pair $Y_{l,ITU}$, $Y_{r,ITU}$. This observation has important consequences for a system that reinstates the binaural parameters of a certain binaural signal pair Y_l, Y_r based on an ITU down-mix $Y_{l,ITU}$, $Y_{r,ITU}$, namely that *no decorrelator is required in the synthesis process*. Instead, it is possible to derive a 2×2 matrix (sub-band-wise, in a similar way as described in Section 8.4) that converts the ITU down-mix to a binaural signal $\hat{Y}_l$, $\hat{Y}_r$ of which the binaural parameters are equal to those of the signal pair Y_l, Y_r resulting from convolution of the original input signals with HRTFs (assuming the HRTF parameter stationarity constraint as outlined in Section 8.3):

$$
\begin{bmatrix} \hat{Y}_l(f) \\ \hat{Y}_r(f) \end{bmatrix} = \mathbf{W}_b \begin{bmatrix} Y_{l,ITU}(f) \\ Y_{r,ITU}(f) \end{bmatrix}
\tag{8.37}
$$

The matrix entries of $\mathbf{W}_b$ now follow from two separate binaural parameter estimation processes. The first column of $\mathbf{W}_b$ represents the binaural parameters stemming from all the original input signals (or virtual sources) present in down-mix signal $Y_{l,ITU}$ (hence, X_{lf}, X_c, and X_{ls}), while the second column of $\mathbf{W}_b$ represents the binaural properties resulting from virtual sources X_{rf}, X_c, and X_{rs}:

$$
\mathbf{W}_b = \begin{bmatrix} \lambda_{11} & \lambda_{12}e^{j\text{-IPD}_R} \\ \lambda_{21}e^{j\text{IPD}_L} & \lambda_{22} \end{bmatrix}
\tag{8.38}
$$

with IPD_L the net IPD resulting from the combined virtual sources X_{lf}, X_c, and X_{ls}, IPD_R the net IPD resulting from X_{rf}, X_c, and X_{rs}, λ_{11} the amplitude ratio between Y_l and $Y_{l,ITU}$, λ_{21} the amplitude ratio between Y_r and $Y_{l,ITU}$, λ_{12} the amplitude ratio between Y_l and $Y_{r,ITU}$, and λ_{22} the amplitude ratio between Y_r and $Y_{r,ITU}$.

8.5 Application to MPEG Surround

8.5.1 Binaural decoding mode

The MPEG Surround parameters aim at faithful perceptual multi-channel reproduction of the original sound stage at the encoder side. In other words, the normal operation mode of MPEG Surround is intended for multi-channel loudspeaker playback. However, the parameters that are transmitted represent identical statistical properties between audio channels to those required for the binaural analysis and synthesis approach described in Sections 8.3 and 8.4. It is therefore possible to integrate the binaural analysis and synthesis approach in a dedicated MPEG Surround *binaural decoding mode* for headphone playback. The architecture of this mode is visualized in Figure 8.2. Instead of directly applying the transmitted spatial parameters to the output signals to generate multi-channel output, the parameters are used in a binaural parameter analysis stage to compute binaural parameters that would result from the combined spatial decoding and binaural rendering process. Thus, the binaural parameter analysis stage estimates the binaural parameters p_{y_l},

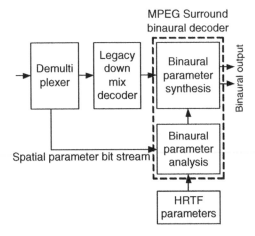

Figure 8.2 Overview of a binaural decoding mode for MPEG Surround.

Figure 8.3   Overview of a binaural synthesis stage based on a mono down-mix.

p_{y_r}, IPD and IC for each parameter band and each newly transmitted parameter set. The binaural output signals are subsequently synthesized by the binaural parameter synthesis stage.

8.5.2 Binaural parameter synthesis

The binaural synthesis process, of which the essence is described in Section 8.4, is performed in a filterbank domain to enable independent processing of various time–frequency tiles. To facilitate integration of the binaural analysis and synthesis method in the MPEG Surround system, the same hybrid quadrature mirror filter (QMF) bank and decorrelators are applied that are currently being used in MPEG Surround. The synthesis stage is outlined in Figure 8.3. A hybrid QMF filterbank provides 71 down-sampled, nonlinearly spaced sub-bands that can be grouped in 28 parameter bands that approximate the bandwidth of critical bands. In case of a mono down-mix, the hybrid-QMF-domain signal is processed by a decorrelator that consists of a lattice all-pass filters to generate a signal that is statistically independent from its input [79, 235]. In case of a stereo down-mix, the two down-mix signals serve as input to the spatial synthesis stage (without decorrelator). Subsequently, a 2×2 matrix $\mathbf{W}_b$ is applied for each parameter band to generate two signals. The final binaural output is obtained by two hybrid QMF synthesis filterbanks.

The 2×2 binaural synthesis matrix $\mathbf{W}_b$ is computed for each received spatial parameter set. These spatial parameter sets are defined for specific temporal positions that are signaled in the MPEG Surround bitstream. For audio samples of positions in between such parameter positions, the synthesis matrix is interpolated linearly.

Thus, apart from the summation of statistical properties across virtual sound sources i given in Equations (8.20–8.22), the rendering method is independent of the number of sound sources. In other words, the parameter analysis stage is the only process of which the complexity depends on the number of virtual sound sources.

8.5.3 Binaural encoding mode

One of the interesting advantages of the parametric approach as described above is that the synthesis process comprises a 2×2 (sub-band) matrix which is, under certain constraints, *invertible*. In other words, the inverse of the 2×2 processing matrix that converts a conventional stereo down-mix to a binaural stereo signal can be employed to recover a conventional stereo down-mix from a binaural stereo signal:

$$\begin{bmatrix} Y_{l,ITU}(f) \\ Y_{r,ITU}(f) \end{bmatrix} = \mathbf{W}_b^{-1} \begin{bmatrix} \hat{Y}_l(f) \\ \hat{Y}_r(f) \end{bmatrix} \qquad (8.39)$$

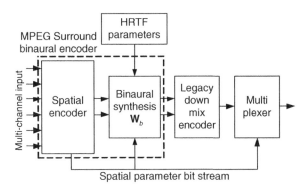

Figure 8.4   MPEG Surround encoder extended with a binaural synthesis stage ('binaural encoder').

This property provides means to move the binaural synthesis process to the MPEG Surround *encoder* as outlined in Figure 8.4. The spatial encoder within the MPEG Surround encoder generates a conventional stereo down-mix from the multi-channel input, accompanied by spatial parameters. The stereo down-mix is subsequently processed using the binaural synthesis matrix $\mathbf{W}_b$. This synthesis matrix is controlled by spatial parameters and HRTF parameters, as described previously. The resulting stereo binaural signal is encoded using the legacy down-mix encoder. With this approach, binaurally rendered audio is provided to legacy decoders.

In case of stereo or multi-channel loudspeaker playback at the receiver side, the corresponding MPEG Surround decoder needs to be extended with a binaural inversion stage that applies matrix $\mathbf{W}_b^{-1}$ on the binaural down mix that reconstructs the conventional stereo down mix. This extension is visualized in Figure 8.5. This scheme works under the constraints that: (1) the legacy stereo audio coder is (to a large extend) waveform preserving; (2) the same HRTF parameters are available at the encoder and the decoder; and (3) special care is taken to ensure nonsingular encoding matrices.

This scheme has the advantage that legacy stereo decoders (for example on mobile devices) will render a virtual multi-channel experience, while MPEG Surround decoders are still capable of decoding high-quality multi-channel signals for loudspeaker playback despite the fact that the transmitted stereo down-mix is heavily altered by the binaural

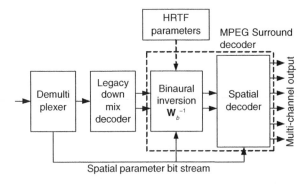

Figure 8.5   MPEG Surround decoder extended with a binaural inversion stage.

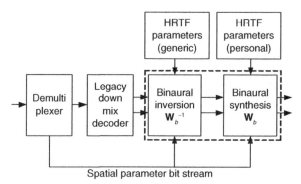

Figure 8.6   Conversion of a binaural signal using nonindividualized (generic) HRTFs to a binaural signal using individual HRTFs.

synthesis process. The drawbacks of the encoder-side binaural synthesis scenario are the need to align encoder and decoder-side HRTF data (for example by including the HRTF parameters in the MPEG Surround bit stream), and the difficulty to use personalized HRTFs. The latter can be accounted for however by cascading the inversion matrix of Fig 8.5 based on the same (generic) HRTFs as applied in the encoder, with an additional synthesis stage based on different, personalized HRTFs. This scheme is depicted in Figure 8.6. Hence, if personalized HRTFs are available at the decoder, the structure of Figure 8.6 is capable of synthesizing a binaural signal using these personalized HRTFs, even if the received down-mix signal was generated using nonpersonalized (generic) HRTFs.

8.5.4 Evaluation

Procedure

A listening test was pursued to evaluate the subjective quality of the proposed binaural synthesis method. In this test, the quality of the MPEG Surround binaural decoding mode (MPS binaural) is compared with a reference condition. This reference condition comprised convolution of an original multi-channel audio excerpt with HRIRs. As a control condition, the combination of MPEG Surround multi-channel decoding followed by conventional HRIR convolution was employed (denoted MPS + HRIR). For all configurations, anechoic KEMAR HRIRs [91] were used with a length of 128 samples at a sampling frequency of 44.1 kHz.

For both the binaural decoding mode as well as the control condition, the same MPEG Surround bitstream was employed. This bitstream was generated using a state-of-the-art MPEG Surround decoder using a mono down-mix configuration. This mono down-mix was subsequently encoded using a high-efficiency AAC encoder at 44 kbps. The spatial parameters generated by the MPEG Surround decoder occupied approximately 4 kbps. This rather low bitrate of 48 kbps total was selected because it is foreseen that the binaural decoding mode is especially suitable for mobile applications that are under severe transmission bandwidth and complexity constraints.

Table 8.1 Test excerpts.

Excerpt	Name	Category
1	BBC applause	Pathological/ambience
2	ARL applause	Pathological/ambience
3	Chostakovitch	Music
4	Fountain music	Pathological/ambience
5	Glock	Pathological
6	Indie2	Movie sound
7	Jackson1	Music
8	Pops	Music
9	Poulenc	Music
10	Rock concert	Music
11	Stomp	Music (with LFE)

Twelve listeners participated in this experiment. All listeners had significant experience in evaluating audio codecs and were specifically instructed to evaluate the overall quality, consisting of the spatial audio quality as well as any other noticeable artifacts. In a double-blind MUSHRA test [148], the listeners had to rate the perceived quality of several processed excerpts against the original (i.e. unprocessed) excerpts on a 100-point scale with 5 anchors, labeled 'bad', 'poor', 'fair', 'good' and 'excellent'. A hidden reference and the low-pass filtered anchor (reference with a bandwidth limitation of 3.5 kHz) were also included in the test. The subjects could listen to each excerpt as often as they liked and could switch in real time between all versions of each excerpt. The experiment was controlled from a PC with an RME Digi 96/24 sound card using ADAT digital out. Digital-to-analog conversion was provided by an RME ADI-8 DS eight-channel D-to-A converter. Beyerdynamic DT990 headphones were used throughout the test. Subjects were seated in a sound-insulated listening room.

A total of 11 critical excerpts were used as listed in Table 8.1. The excerpts are the same as used in the MPEG Call for Proposals (CfP) on Spatial Audio Coding [142], and range from pathological signals (designed to be critical for the technology at hand) to movie sound and multi-channel music productions. All input and output excerpts were sampled at 44.1 kHz.

Results

The results of the listening test are shown in Figure 8.7. The various excerpts are given along the abscissa, while the ordinate corresponds to the average MUSHRA score across listeners. Different symbols refer to different configurations. The error bars denote the 95% confidence intervals of the means.

The hidden reference (square symbols) has the highest scores. The results for the binaural decoding mode are denoted by the diamonds; the control condition using convolution is represented by the downward triangles. Although the scores for these methods vary between 45 and 85, the binaural decoding approach has scores that are higher than the

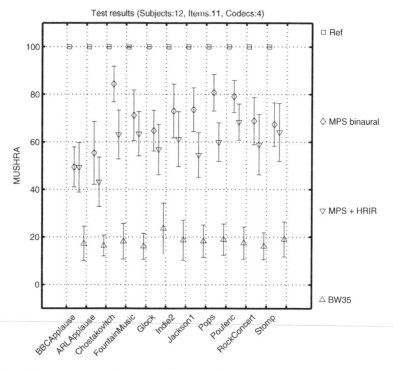

Figure 8.7   Subjective test results averaged across subjects for each excerpt. Error bars denote the 95% confidence intervals of the means.

conventional method for all excerpts. Finally, the low-pass anchor has the lowest scores of around 20.

The average scores for each method across subjects and excerpts are shown in Figure 8.8. The difference between the binaural decoding mode and the control condition amounts to 12 points in favor of the binaural decoder.

If the computational complexity of the binaural decoder and the conventional systems are compared, interesting differences are observed. The number of operations (expressed in multiply-accumulates per second which represent the combined, single processor operation of multiplication followed by addition) amounts to 11.1 million for the binaural decoder and 47 million for the MPEG Surround multi-channel decoder followed by convolution using fast Fourier transforms.

Discussion

The results of the perceptual evaluation indicate that both binaural rendering methods (the binaural decoding mode and the conventional HRIR convolution method) are distinguishable from the reference. This is most probably due to the low bitrate (48 kbps total) that was employed to represent the multi-channel signal in MPEG Surround format. For loudspeaker playback, the perceived quality of MPEG Surround operating at 48 kbps has

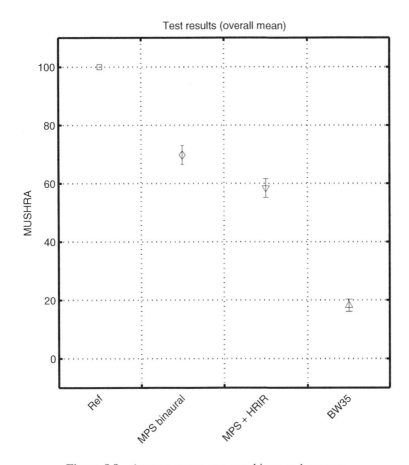

Figure 8.8   Average scores across subjects and excerpts.

been shown to amount 65 in other tests [48, 265]. In that respect, the quality for the test and control conditions are in line with earlier reports.

The parametric representation of MPEG Surround aims at perceptual reconstruction of multi-channel audio. As such, at the bitrate that was under test, MPEG Surround does not deliver full waveform reconstruction of the multi-channel output signals. Such waveform reconstruction requires the use of 'residual coding' as supported by MPEG Surround. However, residual coding results in a significant increase in the bitrate which is undesirable or even unavailable in mobile applications. Given the low scores for MPEG Surround decoding followed by HRIR convolution, the multi-channel signals resulting from the parametric representation seem unsuitable for further post-processing using HRIRs. A high sensitivity for artefacts during post-processing of decoded output is a property that is often observed for lossy audio coders. The binaural decoding mode, however, which does not rely on post-processing of decoded signals, outperforms the convolution-based method, both in terms of perceived quality and computational complexity. This clearly indicates the advantages of parameter-domain processing compared with the signal-domain approach.

8.6 Conclusions

A novel method for binaural rendering based on parametric representations has been outlined. In contrast to conventional, convolution-based methods, head-related impulse responses are transformed to the parameter domain and combined with parameters that describe the statistical properties of the various signals that are radiated by virtual sources. From the combination of statistical properties between virtual source signals and HRTF parameters, *binaural* parameters were derived that describe the relevant perceptual properties at the level of the eardrums. A binaural rendering stage subsequently reinstates the binaural parameters obtained to various time/frequency tiles of a down-mix of the various virtual source signals.

The proposed method can be integrated with parametric multi-channel audio coders that rely on inter-channel cues such as level differences and inter-channel correlations. As such, it is currently an integral part of the MPEG Surround specification, including extensions to stereo down-mixes and support for anechoic HRTFs (BRIRs). Results of a listening test revealed that the proposed method outperforms conventional, convolution-based methods in terms of perceived quality and computational complexity. These properties, combined with the unsurpassed compression efficiency of MPEG Surround, make this MPEG Surround extension very suitable for mobile applications.

9

Audio Coding with Mixing Flexibility at the Decoder Side

9.1 Introduction

In a conventional audio production and transmission chain, the audio content is first produced for playback using a certain reproduction system (for example two-loudspeaker stereophony), and is subsequently encoded, transmitted or stored, and decoded. The specific order of production and encoding/decoding makes it very difficult to enable user interactivity to modify the 'mix' produced by, for example, a studio engineer.

There are however several applications that may benefit from user control in mixing and rendering parameters. For example, in a teleconferencing application, individual users may want to control the spatial position and the loudness of each individual talker. For radio and television broadcasts, users may want to enhance the level of a voice-over for maximum speech intelligibility. Younger people may want to make a 're-mix' of a music track they recently acquired, with control of various instruments and vocals present in the mix.

Conventional 'object-based' audio systems require storage/transmission of the audio sources such that they can be mixed at the decoder side as desired. Also wave field synthesis systems are often driven with audio source signals. ISO/IEC MPEG-4 [139, 146, 230] addresses a general object-based coding scenario. It defines the scene description (= mixing parameters) and uses for each ('natural') source signal a separate mono audio coder. However, when a complex scene with many sources is to be coded the bitrate becomes high since the bitrate scales with the number of sources. An object-based audio system is illustrated in Figure 9.1. A number of audio source signals are coded and stored/transmitted. The receiver mixes the decoded audio source signals to generate stereo [28, 227], surround [150, 227], wavefield synthesis [16, 17, 264], or binaural signals [14, 26] based on mixing parameters.

It would be desirable to have an efficient coding paradigm for audio sources that will be mixed after decoding. However, from an information theoretic point of view, there is no additional coding gain when jointly coding independent sources compared

Spatial Audio Processing: MPEG Surround and Other Applications   Jeroen Breebaart and Christof Faller
© 2007 John Wiley & Sons, Ltd

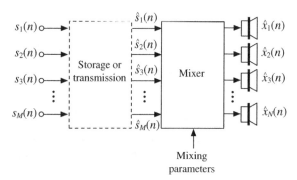

Figure 9.1 The general problem that is addressed – coding of a number of source signals for the purpose of mixing stereo, multi-channel, wavefield synthesis, or binaural audio signals with the decoded source signals.

with independently coding them. For example, given a number of independent instrument signals the best one can do with conventional wisdom is to apply to each instrument signal one coder (e.g. a perceptual audio coder such as AAC [137], AC-3 [88], ATRAC [258], MP3 [136], or PAC [243]).

The scheme outlined in this chapter provides means for joint coding in a more efficient manner than independent coding of individual source signals. Sources are joint-coded for the purpose of mixing after decoding. In this case, considering properties of the source signals, the mixing process, and knowledge on spatial perception it is possible to reduce the bitrate significantly. The source signals are represented as a single mono sum (down-mix) signal plus about 3 kb/s side information per source. Conventional coding would require about $10 \times 80 = 800$ kb/s for 10 sources, assuming a bit budget of 80 kb/s for each source. The described technique requires only about $80 + 10 \times 3 = 110$ kb/s and thus is significantly more efficient than conventional coding.

The scheme for joint-coding of audio source signals is illustrated in Figure 9.2. It is based on coding the *sum* of all audio source signals accompanied by additional parametric side information. This side information represents the statistical properties of the source signals which are the most important factors determining the perceptual spatial cues of the mixer output signals. These properties are temporally evolving spectral envelopes and autocorrelation functions. About 3 kb/s of side information is required per source signal. A conventional audio or speech coder is used to efficiently represent the sum signal.

At the receiving end, the source signals are recovered such that the before mentioned statistical properties approximate the corresponding properties of the original source signals. A stereo, surround, wavefield synthesis, or binaural mixer is applied after decoding of the source signals to generate the output signal.

Conceptually, the aim of the scheme is not to recover the *clean source signals* and it is not intended that one listens to these source signals separately prior to mixing. The goal is that the *mixed output signal* perceptually approximates the reference signal (i.e. the signal generated with the same mixer supplied with the original source signals).

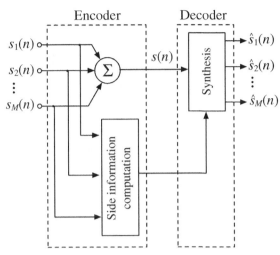

Figure 9.2 A number of source signals are stored or transmitted as sum signal plus information. The side information represents statistical properties of the source signals. At the receiver the source signals are recovered.

9.2 Motivation and details

As mentioned above, the scheme for joint-coding of audio source signals, shown in Figure 9.2, is based on transmission of the sum of the audio source signals,

$$s(n) = \sum_{i=1}^{M} s_i(n) \tag{9.1}$$

where M is the number of source signals and $s_i(n)$ are the individual source signals.

Similar to spatial audio coding techniques, this method relies on the assumption that the perceived auditory spatial image is largely determined by the inter-channel time difference (ICTD), inter-channel level difference (ICLD), and inter-channel coherence (ICC) between the rendered audio channels. Therefore, as opposed to requiring 'clean' source signals $s_i(n)$ as mixer input in Figure 9.1, only signals $\hat{s}_i(n)$ are required that result in similar ICTD, ICLD, and ICC at the mixer output as for the case of supplying the real source signals $s_i(n)$ to the mixer. There are three goals for the generation of $\hat{s}_i(n)$:

- If $\hat{s}_i(n)$ are supplied to a mixer, the mixer output channels will have approximately the same spatial cues (ICLD, ICTD, ICC) as if $s_i(n)$ were supplied to the mixer.

- $\hat{s}_i(n)$ are to be generated with as little as possible information about the original source signals $s_i(n)$ (because the goal is to have low bitrate side information).

- $\hat{s}_i(n)$ are generated from the transmitted sum signal $s(n)$ such that a minimum amount of signal distortion is introduced.

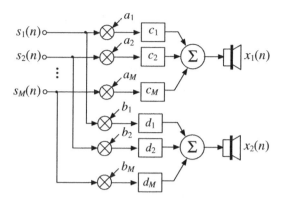

Figure 9.3   A mixer for generating stereo signals given a number of source signals.

Without loss of generality for deriving the scheme for an arbitrary number of rendered output channels, a stereo mixer is considered. A further simplification over the general case is that only amplitude and delay modifications are applied for mixing. (Equalization of objects is also provided by frequency-dependent amplitude changes.) If the discrete source signals were available at the decoder, a stereo signal would be mixed, as shown in Figure 9.3, i.e.

$$x_1(n) = \sum_{i=1}^{M} a_i s_i(n - c_i)$$

$$x_2(n) = \sum_{i} b_i s_i(n - d_i) \tag{9.2}$$

where a_i and b_i represent level mixing parameters, while c_i, and d_i represent time delay mixing parameters.

Given, for each source with index i, the gain G_i (in dB), pan pot position ΔL_i (expressed as level difference in dB), and the delay pan pot position τ_i in samples, the mixing parameters (9.2) can be computed:

$$a_i = \frac{10^{G_i/20}}{\sqrt{1 + 10^{\Delta L_i/10}}}$$

$$b_i = 10^{(G_i + \Delta L_i)/20} a_i$$

$$c_i = \max\{-\tau_i, 0\}$$

$$d_i = \max\{\tau_i, 0\} \tag{9.3}$$

In the following, ICTD, ICLD, and ICC of the stereo mixer output $x_1(n)$, $x_2(n)$ are computed as a function of (statistical properties of) the input source signals $s_i(n)$ and the mixing parameters a_i, b_i, c_i, and d_i. The expressions obtained will give an indication of which source signal properties determine the mixer output ICTD, ICLD, and ICC (together with the mixing parameters).

9.2.1 ICTD, ICLD and ICC of the mixer output

The ICTD, ICLD and ICC parameters of the mixer output are estimated in sub-bands (critical bands) and as a function of time. It is assumed that the source signals $s_i(n)$ are zero mean and mutually independent. A pair of sub-band signals of the mixer output (Equation 9.2) is denoted $\tilde{x}_1(n)$ and $\tilde{x}_2(n)$. Note that for simplicity of notation we are using the same time index n for time-domain and sub-band-domain signals. Also, no sub-band index is used and the analysis/processing described is applied to subbands at each frequency independently. The subband powers of the two mixer output signals are:

$$E\{\tilde{x}_1^2(n)\} = \sum_{i=1}^{M} a_i^2 E\{\tilde{s}_i^2(n)\}$$

$$E\{\tilde{x}_2^2(n)\} = \sum_{i=1}^{M} b_i^2 E\{\tilde{s}_i^2(n)\} \tag{9.4}$$

where $\tilde{s}_i(n)$ is one sub-band signal of source $s_i(n)$ and $E\{.\}$ denotes short-time mean, e.g.

$$E\{\tilde{s}_i^2(n)\} = \frac{1}{K} \sum_{n-K/2}^{n+K/2-1} \tilde{s}_i^2(n) \tag{9.5}$$

where K determines the length of the moving average. (The solutions provided are based on an assumption of incoherent source signals. For partially coherent signals, the reader is referred to Chapter 8, Section 8.3.3.) Note that the sub-band power values $E\{\tilde{s}_i^2(n)\}$ represent for each source signal the spectral envelope as a function of time. The time span considered for the averaging (9.5) determines the time resolution at which the inter-channel cues are considered.

The ICLD, $\Delta L(n)$, between signals $\tilde{x}_1(n), \tilde{x}_2(n)$ is given by:

$$\Delta L(n) = 10 \log_{10} \frac{\sum_{i=1}^{M} b_i^2 E\{\tilde{s}_i^2(n)\}}{\sum_{i=1}^{M} a_i^2 E\{\tilde{s}_i^2(n)\}} \tag{9.6}$$

For estimating ICTD and ICC the normalized cross-correlation function [181]

$$\Phi(n, d) = \frac{E\{\tilde{x}_1(n)\tilde{x}_2(n+d)\}}{\sqrt{E\{\tilde{x}_1^2(n)\}E\{\tilde{x}_2^2(n+d)\}}} \tag{9.7}$$

is computed. The ICC, $c(n)$, is obtained according to

$$c(n) = \max_{d} |\Phi(n, d)| \tag{9.8}$$

For the computation of the ICTD, $\tau(n)$, the location of the highest peak of the cross-correlation function along the delay axis is computed,

$$\tau(n) = \arg \max_{d} \Phi(n, d) \tag{9.9}$$

The normalized cross-correlation function can be computed as a function of the mixing parameters. Together with Equation (9.2), Equation (9.7) can be written as

$$\Phi(n, d) = \frac{\sum_{i=1}^{M} E\{a_i b_i \tilde{s}_i(n - c_i)\tilde{s}_i(n - d_i + d)\}}{\sqrt{E\{\sum_{i=1}^{M} a_i^2 s_i^2(n - c_i)\}E\{\sum_{i=1}^{M} b_i^2 s_i^2(n - d_i)\}}} \tag{9.10}$$

which is equivalent to

$$\Phi(n, d) = \frac{\sum_{i=1}^{M} a_i b_i E\{\tilde{s}_i^2(n)\}\Phi_i(n, d - \tau_i)}{\sqrt{(\sum_{i=1}^{M} a_i^2 E\{\tilde{s}_i^2(n)\})(\sum_{i=1}^{M} b_i^2 E\{\tilde{s}_i^2(n)\})}} \tag{9.11}$$

where the normalized autocorrelation function $\Phi_i(n, e)$ is

$$\Phi_i(n, e) = \frac{E\{s_i(n)s_i(n + e)\}}{E\{s_i^2(n)\}}, \tag{9.12}$$

and $\tau_i = d_i - c_i$. Note that for computing Equation (9.11), given Equation (9.10), it has been assumed that the signals are wide sense stationary within the range of delays considered, i.e.

$$E\{\tilde{s}_i^2(n)\} = E\{\tilde{s}_i^2(n - c_i)\}$$

$$E\{\tilde{s}_i^2(n)\} = E\{\tilde{s}_i^2(n - d_i)\}$$

$$E\{\tilde{s}_i(n)\tilde{s}_i(n + c_i - d_i + d)\} = E\{\tilde{s}_i(n - c_i)\tilde{s}_i(n - d_i + d)\}$$

A numerical example for two source signals, illustrating the dependence between ICTD, ICLD, and ICC and the source sub-band power, is shown in Figure 9.4. The top, middle, and bottom panel of Figure 9.4 show $\Delta L(n)$, $\tau(n)$, and $c(n)$, respectively, as a function of the ratio of the sub-band power of the two sources, $a = E\{\tilde{s}_1^2(n)\}/(E\{\tilde{s}_1^2(n)\} + E\{\tilde{s}_2^2(n)\})$, for different mixing parameters (9.3) ΔL_1, ΔL_2, τ_1, and τ_2 (with $G_i = 1$).

The top panel of Figure 9.4 indicates that when only one source has power in the sub-band ($a = 0$ or $a = 1$), then the mixer ICLD, $\Delta L(n)$ (9.6), is equal to the amplitude panning parameter ΔL_i (Equation 9.3) of the dominant source. When the power in the sub-bands fades from one source to the other, i.e. when a changes from zero to one, the mixer output level difference fades from the amplitude panning parameter of one source to the amplitude panning parameter of the other source.

The middle panel of Figure 9.4 indicates that when only one source has power in the sub-band ($a = 0$ or $a = 1$), then the mixer ICTD, $\tau(n)$ (Equation 9.9), is equal to the delay panning parameter τ_i (Equation 9.3) of the dominant source. As opposed to the mixer output level difference, the mixer output time difference is determined by the delay panning parameter of the source which has more power in the sub-band, as indicated by the hard switch of $\tau(n)$ at $a = 0.5$.

The bottom panel of Figure 9.4 indicates that when only one source has power in the sub-band ($a = 0$ or $a = 1$), then the mixer output coherence, $c(n)$ (9.8), is equal to one. Mixer output coherence decreases when more than one source has power in the sub-band.

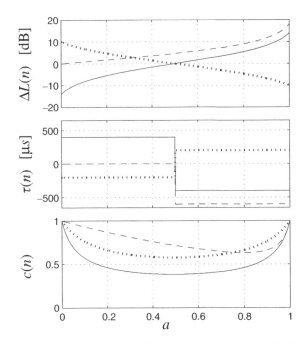

Figure 9.4 $\Delta L(n)$ (top), $\tau(n)$ (middle), and $c(n)$ (bottom) for a critical band as a function of $a = E\{\tilde{s}_1^2(n)\}/(E\{\tilde{s}_1^2(n)\} + E\{\tilde{s}_2^2(n)\})$. The mixer parameters (Equation 9.3) are: $\Delta L_1 = 14$ dB, $\Delta L_2 = -14$ dB, $\tau_1 = -400$ μs, $\tau_2 = 400$ μs (solid); $\Delta L_1 = 18$ dB, $\Delta L_2 = 0$ dB, $\tau_1 = -600$ μs, $\tau_2 = 0$ μs (dashed); $\Delta L_1 = -10$ dB, $\Delta L_2 = 10$ dB, $\tau_1 = 200$ μs, $\tau_2 = -200$ μs (dotted). The source gain has always been chosen to be $G_i = 0$ dB.

9.3 Side information

The previously derived expressions for the inter-channel cues occurring when mixing the (original) source signals indicate which properties determine the inter-channel cues of the mixer output signal. The ICLD (Equation 9.6) depends on the mixing parameters (a_i, b_i) and on the short-time subband power of the sources, $E\{\tilde{s}_i^2(n)\}$ (Equation 9.5). The normalized subband cross-correlation function $\Phi(n, d)$ (Equation 9.11), that is needed for ICTD (Equation 9.9) and ICC (Equation 9.8) computation, depends on $E\{\tilde{s}_i^2(n)\}$ and additionally on the normalized sub-band autocorrelation function, $\Phi_i(n, e)$ (Equation 9.12), for each source signal. If no time adjustments are applied (i.e. $c_i = d_i = 0$), only subband powers are required, without autocorrelation functions. Finally, the level of the mixer output channels depends on the mixing parameters (a_i, b_i) and sub-band power of the sources $E\{\tilde{s}_i^2(n)\}$.

Hence, given the rendering parameters (a_i, b_i, c_i, d_i), the spatial parameters of the rendered scene can be obtained based on the source subband powers $E\{\tilde{s}_i^2(n)\}$ which hence have to be transmitted along with the down-mix signal $s(n)$. In order to reduce the amount of side information, the relative dynamic range of the source signal parameters is limited. At each time, for each sub-band, the power of the strongest source is selected. It was found to suffice to lower bound the corresponding subband power of all the other sources at a value 24 dB lower than the strongest subband power. Thus, the dynamic range of the quantizer can be limited to 24 dB.

The power of the sources with indices $2 \leq i \leq M$ relative to the power of the first source is transmitted as side information,

$$\Delta \tilde{p}_i(n) = 10 \log_{10} \frac{E\{\tilde{s}_i^2(n)\}}{E\{\tilde{s}_1^2(n)\}} \tag{9.13}$$

Note that dynamic range limiting as described previously is carried out prior to Equation (9.13), avoiding numerical problems when $E\{\tilde{s}_1^2(n)\}$ vanishes. A total of 20 sub-bands was used in combination with a parameter update rate for each subband $\Delta \tilde{p}_i(n)$ $(2 \leq i \leq M)$ of about 12 ms. The relative power values are quantized and Huffman coded, resulting in a bitrate of approximately $3(M-1)$ kb/s [84].

9.3.1 Reconstructing the sources

Figure 9.5 illustrates the process that is used to re-create the source signals, given the sum signal (Equation 9.1). This process is part of the 'Synthesis' block in Figure 9.2. The individual source signals are recovered by scaling each sub-band of the sum signal with $g_i(n)$ and by applying a decorrelation filter with impulse response $h_i(n)$,

$$\hat{s}_i(n) = h_i(n) \star (g_i(n)\tilde{s}(n))$$

$$= h_i(n) \star \left(\sqrt{\frac{E\{\tilde{s}_i^2(n)\}}{E\{\tilde{s}^2(n)\}}} \tilde{s}(n) \right) \tag{9.14}$$

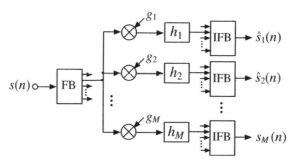

Figure 9.5 The process for generation of $\hat{s}_i(n)$. The sum signal is converted to the sub-band domain. The sub-bands are scaled such that the sub-band power is approximately the same as the sub-band power of the original source signals. Filtering is applied to the scaled sub-bands for decorrelation. The shown processing is carried out independently for each sub-band. FB is a filterbank with sub-bands with bandwidths motivated by perception. IFB is the corresponding inverse filterbank.

where $\star$ is the linear convolution operator and $E\{\tilde{s}_i^2(n)\}$ is computed with the side information by

$$E\{\tilde{s}_i^2(n)\} = \begin{cases} 1/\sqrt{1 + \sum_{i=2}^{M} 10^{\frac{\Delta \tilde{p}_i(n)}{10}}} & \text{for } i = 1 \\ 10^{\frac{\Delta \tilde{p}_i(n)}{10}} E\{\tilde{s}_1^2(n)\}, & \text{otherwise} \end{cases}$$ (9.15)

As decorrelation filters $h_i(n)$, complementary comb filters [179], allpass filters [79, 232], delays [34], or filters with random impulse responses [82, 83] may be used. The goal for the decorrelation process is to reduce correlation between the signals while not modifying how the individual waveforms are perceived. Different decorrelation techniques cause different artifacts. Complementary comb filters cause coloration. All the described techniques are spreading the energy of transients in time causing artifacts such as 'pre-echoes'. Given their potential for artifacts, decorrelation techniques should be applied as little as possible.

When applying no decorrelation processing ($h_i(n) = \delta(n)$ in Equation (9.14)) good audio quality can also be achieved. It is a compromise between artifacts introduced by the decorrelation processing and artifacts due to the fact that the source signals $\hat{s}_i(n)$ are correlated. In fact, it is only beneficial to reconstruct ICC at the mixer *output* rather than the mixer input signals (i.e. the recovered source signals) by integrating the source estimation and mixing processes in the parametric domain, as outlined in the next section.

9.4 Using spatial audio decoders as mixers

Mixing is preferably applied to the transmitted sum signal (Equation 9.1) *without* explicit computation of $\hat{s}_i(n)$. In this approach, the sum signal $s(n)$ is directly transformed to a stereo or multi-channel output signal with the correct spatial attributes. A BCC synthesis stage or MPEG Surround decoder can be used for this purpose. In the following, a stereo reproduction case is considered, but all principles described can be applied for generation of multi-channel audio signals as well.

A stereo BCC synthesis scheme, applied for processing the sum signal (Equation 9.1), is shown in Figure 9.6. The scheme comprises a filterbank (FB), gains (g_1, g_2), delays

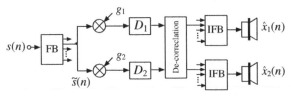

Figure 9.6 A mixer for generating stereo signals directly given the sum of a number of source signals without explicit computation of the source signals. Gain factors, delays, and decorrelation are applied independently in sub-bands.

(D_1, D_2), a decorrelation stage and inverse filterbanks (IFB). The purpose of this scheme is to generate a signal that is perceived similarly as the output signal of a mixer as shown in Figure 9.3. This requires correct synthesis of ICTD, ICLD, ICC and output levels of the BCC output as those obtained when the original source signals were directly fed in to the mixer (see Equation 9.2).

The same side information as for the previously described more general scheme is used, allowing the decoder to have access to the short-time sub-band power values $E\{\tilde{s}_i^2(n)\}$ of the sources. Given $E\{\tilde{s}_i^2(n)\}$, the gain factors g_1 and g_2 in Figure 9.6 are computed as

$$g_1(n) = \sqrt{\frac{\sum_{i=1}^{M} a_i^2 E\{\tilde{s}_i^2(n)\}}{E\{\tilde{s}^2(n)\}}}$$

$$g_2(n) = \sqrt{\frac{\sum_{i=1}^{M} b_i^2 E\{\tilde{s}_i^2(n)\}}{E\{\tilde{s}^2(n)\}}} \qquad (9.16)$$

such that the output subband power and ICLD (9.6) are the same as for the mixer in Figure 9.3.

The ICTD $\tau(n)$ is computed according to Equation (9.9), determining the delays D_1 and D_2 in Figure 9.6,

$$D_1(n) = \max\{-\tau(n), 0\}$$

$$D_2(n) = \max\{\tau(n), 0\} \qquad (9.17)$$

The ICC $c(n)$ is computed according to (9.8) determining the decorrelation processing in Figure 9.6. Decorrelation processing (ICC synthesis) is described in [79, 82, 83, 86, 235] and in Chapters 4, 5 and 6. The advantages of applying decorrelation processing to the mixer output channels compared to applying it for generating independent $\hat{s}_i(n)$ are:

1. Usually the number of source signals M is larger than the number of audio output channels N. Thus, the number of independent audio channels that need to be generated is smaller when de-correlating the N output channels as opposed to de-correlating the M source signals.

2. Often the N audio output channels are correlated (ICC > 0) and less decorrelation processing can be applied than would be needed for generating independent M or N channels.

Due to less decorrelation processing better audio quality is expected.

Best audio quality is obtained when the mixer parameters are constrained such that $a_i^2 + b_i^2 = 1$, i.e. $G_i = 0$ dB. In this case, the sub-band power of each source in the transmitted sum signal (Equation 9.1) is the same as the power of the same source in the mixed decoder output signal. Interestingly, for $G_i = 0$, the decoder output signal (Figure 9.6) is the same as if the mixer output signal (Figure 9.3) were encoded and decoded by a BCC encoder/decoder in this case. Thus, similar audio quality can also be expected.

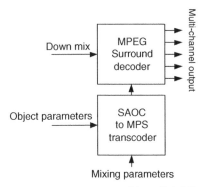

Figure 9.7 Proposed transcoder scheme to convert object (SAOC) parameters to an MPEG Surround (MPS) bit stream based on mixing parameters.

Also, in this case the decoder can not only determine the direction at which each source is to appear, but also the gain of each source can be varied. The gain is increased by choosing $a_i^2 + b_i^2 > 1$ ($G_i > 0$ dB) and decreased by choosing $a_i^2 + b_i^2 < 1$ ($G_i < 0$ dB) in (9.16). If values for G_i are different from zero, it is advisable to constraint the range of G_i between -12 and $+12$ dB.

9.5 Transcoding to MPEG Surround

As described in the previous section, a BCC synthesis engine or MPEG Surround decoder can in principle be used to perform the combined process of source signal decoding and mixing. The MPEG audio subgroup is currently developing a new standard for 'spatial audio object coding' (SAOC) that aims at the combined process of joint source coding and mixing as described in the current chapter. Their approach is to develop a so-called transcoder that generates an MPEG Surround compatible bitstream from transmitted object parameters and mixing parameters (also referred to as 'rendering' parameters). The proposed scheme is outlined in Figure 9.7.

The use of an MPEG Surround decoder as rendering engine has several advantages. First, it provides a proven, and flexible rendering engine for many different output channel configurations. Second, the down-mix may comprise two channels as well. Third, it provides binaural rendering facilities for headphones. But besides the functionality provided by the MPEG Surround decoder, additional functionality is envisioned as well. The proposed system also comprises a pre-processing stage to send of one or more objects to external effect stages, such as reverberators, delays, modulation effects, dynamic range reduction, etc. (not shown in Figure 9.7).

9.6 Conclusions

A scheme for joint object coding and mixing was motivated and derived using a stereo mixer with amplitude panning, delay panning, and gains for each source. Note that from a statistical viewpoint, the side information that represents the relative power in sub-bands between the sources (Equation 9.13), is similar to the ICLD used by a BCC (or MPEG Surround) scheme. However, the usage of these parameters is different. In the case of the joint-source coding scheme, the parameters are relative sub-band power values for each source signal without any spatial characteristic. In the case of BCC, the parameters are the inter-channel cues between audio channels that describe the spatial image of a complete auditory scene.

Also, the joint-coding synthesis stage, shown in Figure 9.5, at first sight looks similar to a BCC synthesis scheme. The differences are that the filters do not have the purpose of reproducing an ICC parameter related to a multi-channel signal, but merely the purpose of mimicking the mutual (time invariant) independence of the output source signals. Another important difference to BCC is that the joint-source coding output signals are not intended for listening in isolation but require post-mixing.

10

Multi-loudspeaker Playback of Stereo Signals

10.1 Introduction

In the previous chapters, coding and rendering methods have been discussed for audio content that was in most cases produced for playback on a predetermined reproduction system, i.e. either stereo playback, a 5.1 home cinema setup, or alike. However, especially in the music domain, audio tracks have predominantly been produced for stereo playback. With the advent of multi-channel audio setups, many consumers would like to engage their full audio reproduction setup, even if the content was produced for two-channel stereophony. Therefore there is a demand for so-called upmix algorithms that convert conventional stereo material to a format for multi-channel playback, either using a standard 5.1 setup, or on more advanced reproduction systems (for example involving wavefield synthesis methods).

In this chapter, a technique is described to generate any number of independent audio channels, given two-channel stereo audio signals [85]. The technique is based on a perceptually motivated spatial decomposition for two-channel stereo audio signals, capturing the information of the virtual sound stage. The spatial decomposition allows to re-synthesize audio signals for playback over other sound systems than two-channel stereo. With the use of more front loudspeakers, the width of the virtual sound stage can be increased beyond $\pm 30°$ and the sweet spot region is extended. Optionally, lateral independent sound components can be played back separately over loudspeakers on the sides of a listener to increase listener envelopment. It is also explained how the spatial decomposition can be used with surround sound and wavefield synthesis based audio systems.

10.2 Multi-channel stereo

Many innovations beyond two-channel stereo have failed because of cost, impracticability (e.g. number of loudspeakers), and last but not least a requirement for backwards compatibility. While 5.1 surround multi-channel audio systems [150, 228] are being adopted

Spatial Audio Processing: MPEG Surround and Other Applications   Jeroen Breebaart and Christof Faller
© 2007 John Wiley & Sons, Ltd

widely by consumers, this system is compromised in terms of number of loudspeakers and with a backwards compatibility restriction (the front left and right loudspeakers are located at the same angles as in two-channel stereo, i.e. ±30°, resulting in a narrow frontal virtual sound stage).

It is a fact that by far most audio content is available in the two-channel stereo format. (In the following the term *stereo* is used for two-channel stereo.) For audio systems enhancing the sound experience beyond stereo, it is thus crucial that stereo audio content can be played back, desirably with an improved experience compared with the legacy systems.

It has long been realized that the use of more front loudspeakers improves the virtual sound stage, especially for listeners not exactly located in the sweet spot [169, 255]. Therefore, there has been a lot of attention on playing back stereo signals with an additional center loudspeaker [164, 246] (Reprinted in [74],) [9, 94]. Gerzon [94] also treats the more general case of up-mixing signals. However, the improvement of these techniques over conventional stereo playback has not been sufficient for mass adoption. The main limitations of these techniques are that they consider only localization and not explicitly other aspects such as ambience and listener envelopment. Further, the localization theory behind these techniques is based on a one-virtual-source scenario, limiting their performance when a number of sources are present at different directions simultaneously.

These weaknesses are resolved by the described technique using a perceptually motivated spatial decomposition of stereo audio signals. Given this decomposition, audio signals are rendered for an increased number of loudspeakers, loudspeaker line arrays, and wavefield synthesis systems [17, 264].

10.3 Spatial decomposition of stereo signals

Stereo signals are recorded or mixed such that for each source the signal goes coherently into the left and right signal channel with specific directional cues (level difference, time difference) and reflected/reverberated independent signals go into the channels determining auditory object width and listener envelopment cues. This motivates modeling single source stereo signals, as illustrated in Figure 10.1, where the signal s mimics the direct sound from a direction determined by the factor a. The independent signals, n_1 and n_2, correspond to the lateral reflections. These signals are assumed to have the following relation with the stereo signal pair x_1, x_2:

$$x_1(k) = s(k) + n_1(k)$$
$$x_2(k) = as(k) + n_2(k) \tag{10.1}$$

In order to get a decomposition which is not only effective in a one auditory object scenario, but nonstationary scenarios with multiple concurrently active sources, the described decomposition is carried out independently in a number of frequency bands and adaptively in time:

$$x_{1,m}(k) = s_m(k) + n_{1,m}(k)$$
$$x_{2,m}(k) = A_b s_m(k) + n_{2,m}(k) \tag{10.2}$$

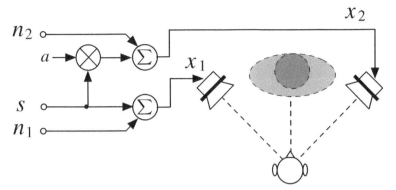

Figure 10.1   Mixing a stereo signal mimicking direct sound s and lateral reflections n_1 and n_2. The factor A determines the direction at which the auditory object appears.

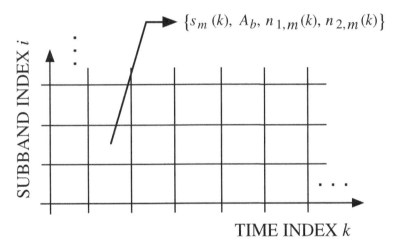

Figure 10.2   Each left and right time–frequency tile of the stereo signal, x_1 and x_2, is decomposed into three signals, s, n_1, and n_2, and a factor A.

where m is the sub-band index, k is the time index, and A_b the amplitude factor for signal s_m for a certain parameter band b that may comprise one or more sub-bands of the sub-band signals (see also Chapter 5 for more details on sub-bands and parameter bands). The decomposition in separate time/frequency tiles is illustrated in Figure 10.2, i.e. in each time–frequency tile with indices m and k, the signals s_m, $n_{1,m}$, $n_{2,m}$, and factor A_b are estimated independently. For brevity of notation, the sub-band and time indices are often ignored in the following. Similarly as BCC or MPEG Surround a perceptually motivated sub-band decomposition is used. This decomposition may be based on the fast fourier transform, quadrature mirror filterbank, or other filterbank. For each parameter band, the signals s_m, $n_{1,m}$, $n_{2,m}$, and A_b are estimated based on segments with a length of approximately 20 ms.

Given the stereo sub-band signal pair, $x_{1,m}$ and $x_{2,m}$, the goal is to estimate s_m, A_b, $n_{1,m}$, and $n_{2,m}$ in each parameter band. This is performed by analysis of the powers and cross-correlation of the stereo signal pair. A short-time estimate of the power of $x_{1,m}$ in parameter band b is denoted $p_{x_1,b}$ and is obtained as outlined in Chapter 6, Section 6.3.4. The powers of $n_{1,m}$ and $n_{2,m}$ in each parameter band are assumed to be the same, i.e. it is assumed that the amount of lateral independent sound is the same for the left and right signals:

$$p_{n_1,b} = p_{n_2,b} = p_{n,b} \tag{10.3}$$

10.3.1 Estimating $p_{s,b}$, A_b and $p_{n,b}$

Given the sub-band representation of the stereo signal, the power ($p_{x_1,b}$, $p_{x_2,b}$) and the normalized cross-correlation $\rho_{x_1x_2,b}$ for parameter band b are computed (see also Section 6.3.4). A_b, $p_{s,b}$, and $p_{n,b}$ are subsequently estimated as a function of the estimated $p_{x_1,b}$, $p_{x_2,b}$, and $\rho_{x_1x_2,b}$. Three equations relating the known and unknown variables are:

$$p_{x_1,b} = p_{s,b} + p_{n,b}$$

$$p_{x_2,b} = A_b^2 p_{s,b} + p_{n,b}$$

$$\rho_{x_1x_2,b} = \frac{A_b p_{s,b}}{\sqrt{p_{x_1,b} p_{x_2,b}}} \tag{10.4}$$

These equations solved for A_b, $p_{s,b}$, and $p_{n,b}$, yield

$$A_b = \frac{B_b}{2C_b}$$

$$p_{s,b} = \frac{2C_b^2}{B_b}$$

$$p_{n,b} = p_{x_1,b} - \frac{2C_b^2}{B_b} \tag{10.5}$$

with

$$B_b = p_{x_2,b} - p_{x_1,b} + \sqrt{(p_{x_1,b} - p_{x_2,b})^2 + 4p_{x_1,b} p_{x_2,b} \rho_{x_1x_2,b}^2}$$

$$C_b = \rho_{x_1x_2,b} \sqrt{p_{x_1,b} p_{x_2,b}} \tag{10.6}$$

10.3.2 Least-squares estimation of s_m, $n_{1,m}$ and $n_{2,m}$

Next, the least-squares estimates of s_m, $n_{1,m}$, and $n_{2,m}$ are computed as a function of A_b, $p_{s,b}$, and $p_{n,b}$. For each parameter band b and each independent signal frame, the signal s_m is estimated as

$$\hat{s}_m(k) = w_{1,b} x_{1,m}(k) + w_{2,b} x_{1,m}(k)$$

$$= w_{1,b}(s_m(k) + n_{1,m}(k)) + w_{2,b}(A_b s_m(k) + n_{2,m}(k)) \tag{10.7}$$

where $w_{1,b}$ and $w_{2,b}$ are real-valued weights. The estimation error is

$$E(k) = (1 - w_{1,b} - w_{2,b}A_b)s_m(k) - w_{1,b}n_{1,m}(k) - w_{2,b}n_{2,m}(k) \qquad (10.8)$$

The weights $w_{1,b}$ and $w_{2,b}$ are optimal in a least mean-square sense when the error signal E is orthogonal to $x_{1,m}$ and $x_{2,m}$ in parameter band b [117], i.e.

$$\rho_{Ex_1,b} = 0$$
$$\rho_{Ex_2,b} = 0 \qquad (10.9)$$

yielding two equations:

$$(1 - w_{1,b} - w_{2,b}A_b)p_{s,b} - w_{1,b}p_{n,b} = 0$$
$$A_b(1 - w_{1,b} - w_{2,b}A_b)p_{s,b} - w_{2,b}p_{n,b} = 0 \qquad (10.10)$$

from which the weights are computed:

$$w_{1,b} = \frac{p_{s,b}p_{n,b}}{(A_b^2 + 1)p_{s,b}p_{n,b} + p_{n,b}^2}$$

$$w_{2,b} = \frac{A_b p_{s,b}p_{n,b}}{(A_b^2 + 1)p_{s,b}p_{n,b} + p_{n,b}^2} \qquad (10.11)$$

Similarly, $n_{1,m}$ and $n_{2,m}$ are estimated. The estimate of $n_{1,m}$ is

$$\hat{n}_{1,m}(k) = w_{3,b}x_{1,m}(k) + w_{4,b}x_{2,m}(k)$$
$$= w_{3,b}(s_m(k) + n_{1,m}(k)) + w_{4,b}(A_b s_m(k) + n_{2,m}(k)) \qquad (10.12)$$

The estimation error is

$$E(k) = (-w_{3,b} - w_{4,b}A_b)s_m(k) - (1 - w_{3,b})n_{1,m}(k) - w_{2,b}n_{2,m}(k) \qquad (10.13)$$

Again, the weights are computed such that the estimation error is orthogonal to $x_{1,m}$ and $x_{2,m}$, resulting in

$$w_{3,b} = \frac{A_b^2 p_{s,b}p_{n,b} + p_{n,b}^2}{(A_b^2 + 1)p_{s,b}p_{n,b} + p_{n,b}^2}$$

$$w_{4,b} = \frac{-A_b p_{s,b}p_{n,b}}{(A_b^2 + 1)p_{s,b}p_{n,b} + p_{n,b}^2} \qquad (10.14)$$

The weights for computing the least-squares estimate of $n_{2,m}$

$$\hat{n}_{2,m}(k) = w_{5,b}x_{1,m}(k) + w_{6,b}x_{2,m}(k)$$
$$= w_{5,b}(s_m(k) + n_{1,m}(k)) + w_{6,b}(A_b s_m(k) + n_{2,m}(k)) \qquad (10.15)$$

are

$$w_{5,b} = \frac{-A_b p_{s,b} p_{n,b}}{(A_b^2 + 1) p_{s,b} p_{n,b} + p_{n,b}^2}$$

$$w_{6,b} = \frac{p_{s,b} p_{n,b} + p_{n,b}^2}{(A_b^2 + 1) p_{s,b} p_{n,b} + p_{n,b}^2} \tag{10.16}$$

10.3.3 Post-scaling

Given the initial least-squares estimates $\hat{s}_m$, $\hat{n}_{1,m}$, and $\hat{n}_{2,m}$, post-scaling is applied such that the power of the estimates $\hat{s}_m$, $\hat{n}_{1,m}$, and $\hat{n}_{2,m}$ in each parameter band equals to $p_{s,b}$ and $p_{n,b}$. The power of $\hat{s}_m$ in parameter band b is

$$p_{\hat{s},b} = (w_{1,b} + A_b w_{2,b})^2 p_{s,b} + (w_{1,b}^2 + w_{2,b}^2) p_{n,b} \tag{10.17}$$

Thus, for obtaining an estimate of $\hat{s}_m$ with power $p_{s,b}$, $\hat{s}_m$ is scaled

$$\hat{s}'_m(k) = \frac{\sqrt{p_{s,b}}}{\sqrt{(w_{1,b} + A_b w_{2,b})^2 p_{s,b} + (w_{1,b}^2 + w_{2,b}^2) p_{n,b}}} \hat{s}_m(k) \tag{10.18}$$

With similar reasoning, $\hat{n}_{1,m}$ and $\hat{n}_{2,m}$ are scaled, i.e.

$$\hat{n}'_{1,m}(k) = \frac{\sqrt{p_{n,b}}}{\sqrt{(w_{3,b} + A_b w_{4,b})^2 p_{s,b} + (w_{3,b}^2 + w_{4,b}^2) p_{n,b}}} \hat{n}_{1,m}(k)$$

$$\hat{n}'_{2,m}(k) = \frac{\sqrt{p_{n,b}}}{\sqrt{(w_{5,b} + A_b w_{6,b})^2 p_{s,b} + (w_{5,b}^2 + w_{6,b}^2) p_{n,b}}} \hat{n}_{2,m}(k) \tag{10.19}$$

10.3.4 Numerical examples

The factor A_b (top panel), the ratio $p_{s,b}/p_{x_1,b}$ (middle panel) and $A_b^2 p_{s,b}/p_{x_2,b}$ (lower panel) expressed in dB are shown in Figure 10.3 as a function of the stereo level difference $p_{x_2,b}/p_{x_1,b}$ (in dB) and the cross correlation $\rho_{x_1 x_2,b}$.

The weights $w_{1,b}$ and $w_{2,b}$ for computing the least-squares estimate of s_m are shown in the top two panels of Figure 10.4 as a function of the stereo signal level difference and $\rho_{x_1 x_2,b}$. The post-scaling factor for $\hat{s}_m$ (10.18) is shown in the bottom panel.

The weights $w_{3,b}$ and $w_{4,b}$ for computing the least-squares estimate of $n_{1,m}$ and the corresponding post-scaling factor (10.19) are shown in Figure 10.5 as a function of the stereo signal level difference and $\rho_{x_1 x_2,b}$.

The weights $w_{5,b}$ and $w_{6,b}$ for computing the least-squares estimate of $n_{2,m}$ and the corresponding post-scaling factor (10.19) are shown in Figure 10.6 as a function of the stereo signal level difference and $\rho_{x_1 x_2,b}$.

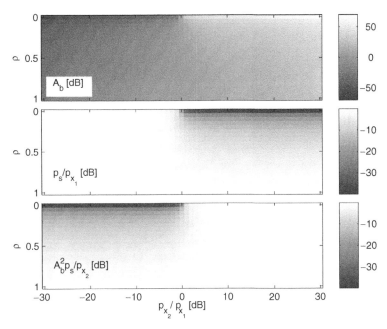

Figure 10.3   The factor A_b (top panel), the ratio $p_{s,b}/p_{x_1,b}$ (middle panel) and $A_b^2 p_{s,b}/p_{x_2,b}$ (lower panel) expressed in dB as a function of the stereo level difference $p_{x_2,b}/p_{x_1,b}$ (in dB) and the cross correlation $\rho_{x_1 x_2,b}$.

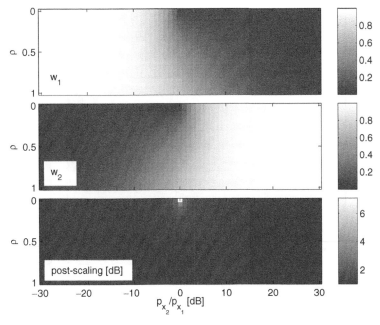

Figure 10.4   The least-squares estimate weights $w_{1,b}$ and $w_{2,b}$ and the post-scaling factor for computation of the estimate of s_m.

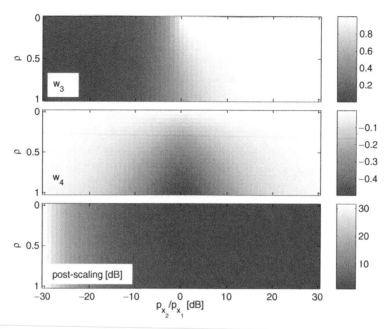

Figure 10.5 The least-squares estimate weights $w_{3,b}$ and $w_{4,b}$ and the post-scaling factor for computation of the estimate of $n_{1,m}$.

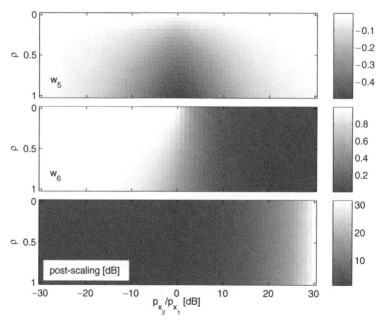

Figure 10.6 The least-squares estimate weights $w_{5,b}$ and $w_{6,b}$ and the post-scaling factor for computation of the estimate of $n_{2,m}$.

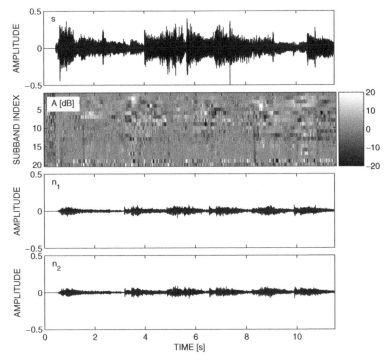

Figure 10.7 Estimates $\hat{s}$, A_b, $\hat{n}_1$, and $\hat{n}_2$ are shown as a function of time for a short audio clip. The factor A_b is shown for various parameter bands.

An example for the spatial decomposition of a stereo rock music clip with a singer in the center is shown in Figure 10.7. The estimates of s, A, n_1, and n_2 are shown. The signals are shown in the time domain (i.e. after independent processing of each parameter band and subsequently transforming the signals to the time domain) and A_b is shown for every time-frequency tile. The estimated direct sound s is relatively strong compared to the independent lateral sound n_1 and n_2 since the singer in the center is dominant.

10.4 Reproduction using different rendering setups

Given the spatial decomposition of the stereo signal, i.e. the sub-band signals for the estimated localized direct sound $\hat{s}'_m$, the factor A_b, and the lateral independent signals $\hat{n}'_{1,m}$ and $\hat{n}'_{2,m}$, one can define rules on how to emit the signal components corresponding to $\hat{s}'_m$, $\hat{n}'_{1,m}$, and $\hat{n}'_{2,m}$ from different playback setups.

10.4.1 Multiple loudspeakers in front of the listener

Figure 10.8 illustrates the scenario that is addressed. The virtual sound stage of width $\phi_0 = 30°$, shown in Part (a) of the figure, is scaled to a virtual sound stage of width ϕ'_0 which is reproduced with multiple loudspeakers, shown in part (b) of the figure.

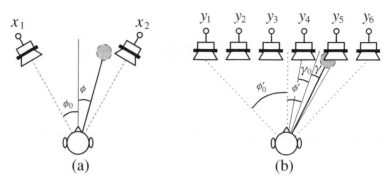

Figure 10.8 The ±30° virtual sound stage (a) is converted to a virtual sound stage with the width of the aperture of a loudspeaker array (b).

The estimated independent lateral sound, $\hat{n}'_{1,m}$ and $\hat{n}'_{2,m}$, is emitted from the loudspeakers on the sides, e.g. loudspeakers 1 and 6 in Figure 10.8(b). That is, because the more the lateral sound is emitted from the side the more it is effective in terms enveloping the listener into the sound [201]. Given the estimated factor A_b, the angle ϕ_b of the auditory object in parameter band b relative to the $\pm\phi_0$ virtual sound stage is estimated, using the 'stereophonic law of sines' [11],

$$\phi_b = \sin^{-1}\left(\frac{A_b - 1}{A_b + 1}\sin\phi_0\right) \tag{10.20}$$

Alternatively, other panning laws, such as the stereophonic law of tangents, may be used. The obtained angle is linearly scaled to compute the angle relative to the widened sound stage,

$$\phi'_b = \frac{\phi'_0}{\phi_0}\phi_b \tag{10.21}$$

The loudspeaker pair enclosing ϕ'_b is selected. In the example illustrated in Figure 10.8(b) this pair has indices 4 and 5. The angles relevant for amplitude panning between this loudspeaker pair, γ_0 and γ, are defined as shown in the figure. If the selected loudspeaker pair has indices l and $l + 1$ then the signals given to these loudspeakers are

$$a_{1,b}\sqrt{1 + A_b^2}s_m(k)$$

$$a_{2,b}\sqrt{1 + A_b^2}s_m(k) \tag{10.22}$$

where the amplitude panning factors $a_{1,b}$ and $a_{2,b}$ are computed with the stereophonic law of sines and normalized such that $a_{1,b}^2 + a_{2,b}^2 = 1$

$$a_{1,b} = \frac{1}{\sqrt{1 + C_b^2}}$$

$$a_{2,b} = \frac{C_b}{\sqrt{1 + C_b^2}} \tag{10.23}$$

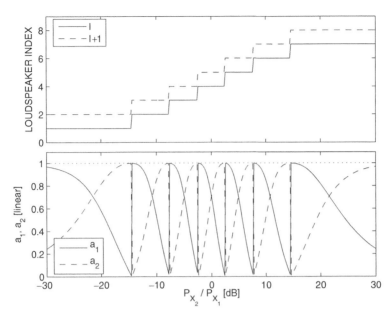

Figure 10.9 Loudspeaker pair selection l (top) and factors $a_{1,b}$ and $a_{2,b}$ are shown as a function of the stereo signal level difference $p_{x_2,b}/p_{x_1,b}$.

with

$$C_b = \frac{\sin(\gamma_0 + \gamma)}{\sin(\gamma_0 - \gamma)} \tag{10.24}$$

The factors $\sqrt{1 + A_b^2}$ in Equation (10.22) are such that the total power of these signals is equal to the total power of the coherent components, s_m and $A_b s_m$, in the stereo signal.

Figure 10.9 shows an example for the selection of loudspeaker pairs, l and $l + 1$, and the amplitude panning factors $a_{1,b}$ and $a_{2,b}$ for $\phi_0' = \phi_0 = 30°$ for $I = 8$ loudspeakers at angles $\{-30°, -20°, -12°, -4°, 4°, 12°, 20°, 30°\}$.

Given the above reasoning, each time–frequency tile of the output signal channels is computed as

$$y_{i,m}(k) = \delta(i - 1)\hat{n}'_{1,m}(k) + \delta(i - I)\hat{n}'_{2,m}(k) +$$

$$(\delta(i - l)a_{1,b} + \delta(i - l - 1)a_{2,b})\sqrt{1 + A_b^2}\hat{s}'_m(k) \tag{10.25}$$

where

$$\delta(i) = \begin{cases} 1 & \text{for } i = 0 \\ 0 & \text{otherwise} \end{cases} \tag{10.26}$$

and i is the output channel index $1 \leq i \leq I$. The sub-band signals of the output channels are converted back to the time domain and form the output channels y_1 to y_I. In the following, this last step is not always explicitly mentioned again.

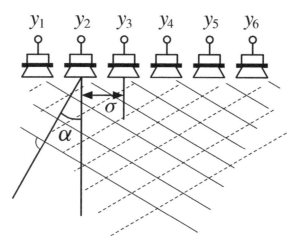

Figure 10.10   Alternatively, $\hat{h}'_1$ and $\hat{h}'_2$ are emitted as two plane waves emitted with angles $\pm\alpha$.

A limitation of the scheme described is that when the listener is at one side, e.g. close to loudspeaker 1, the lateral independent sound will reach him with much more intensity than the lateral sound from the other side. This problem can be circumvented by emitting the lateral independent sound from all loudspeakers with the aim of generating two lateral plane waves. This is illustrated in Figure 10.10. The lateral independent sound is given to all loudspeakers with delays mimicking a plane wave with a certain direction,

$$y_{i,m}(k) = \frac{\hat{n}'_{1,m}(k - (i-1)d_b)}{\sqrt{I}} + \frac{\hat{n}'_{2,m}(k - (I-i)d_b)}{\sqrt{I}} +$$

$$(\delta(i-l)a_{1,b} + \delta(i-l-1)a_{2,b})\sqrt{1 + A_b^2}\hat{s}'_m(k)$$

$$(10.27)$$

where d_b is the delay,

$$d_b = \frac{\sigma f_s \sin\alpha_b}{v} \qquad\qquad (10.28)$$

σ is the distance between the equally spaced loudspeakers (in meters), v is the speed of sound (in meters per second), f_s is the sub-band sampling frequency (in Hertz), and $\pm\alpha_b$ are the directions of propagation of the two plane waves.

10.4.2 Multiple front loudspeakers plus side loudspeakers

The previously described playback scenario aims at widening the virtual sound stage and at making the perceived sound stage independent of the location of the listener.

Optionally one can play back the independent lateral sound, $\hat{n}'_{1,m}$ and $\hat{n}'_{2,m}$, with two separate loudspeakers located more to the sides of the listener, as illustrated in Figure

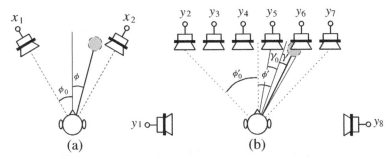

Figure 10.11   The $\pm 30°$ virtual sound stage (a) is converted to a virtual sound stage with the width of the aperture of a loudspeaker array (b). Additionally, the lateral independent sound is played from the sides with separate loudspeakers for a stronger listener envelopment.

10.11. It is expected that this results in a stronger impression of listener envelopment. In this case, the output signals are also computed by Equation (10.25), where the signals with index 1 and I are the loudspeakers on the side. The loudspeaker pair selection, l and $l + 1$, is in this case such that $\hat{s}'_m$ is never given to the signals with index 1 and I since the whole width of the virtual stage is projected to only the front loudspeakers $2 \leq i \leq I - 1$.

Figure 10.12 shows an example for the eight signals generated for the setup shown in Figure 10.11 for the same music clip for which the spatial decomposition was shown in Figure 10.7. Note that the dominant singer in the center is amplitude panned between the center two loudspeaker signals, y_4 and y_5.

10.4.3 Conventional 5.1 surround loudspeaker setup

One possibility to convert a stereo signal to a 5.1 surround compatible multi-channel audio signal is to use a setup as shown in Figure 10.11(b) with three front loudspeakers and two rear loudspeakers arranged as specified in the 5.1 standard. In this case, the rear loudspeakers emit the independent lateral sound, while the front loudspeakers are used to reproduce the virtual sound stage. Informal listening indicates that, when playing back audio signals as described, listener envelopment is more pronounced compared with stereo playback.

Another possibility to convert a stereo signal to a 5.1 surround compatible signal is to use a setup as shown in Figure 10.8 where the loudspeakers are rearranged to match a 5.1 configuration. In this case, the $\pm 30°$ virtual stage is extended to a $\pm 110°$ virtual stage surrounding the listener.

10.4.4 Wavefield synthesis playback system

First, signals are generated similar as for a setup as is illustrated in Figure 10.11(b). Then, I virtual sources are defined in the wavefield synthesis system. The lateral independent sound, y_1 and y_I, is emitted as plane waves or sources in the far field as is illustrated in

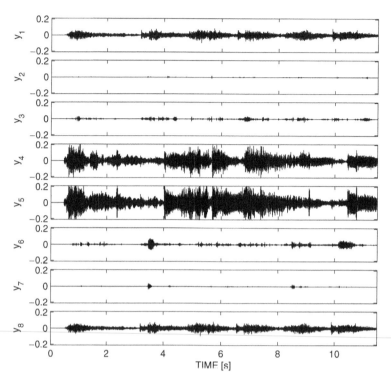

Figure 10.12   The eight signals, generated for a setup as in Figure 10.11(b), are shown.

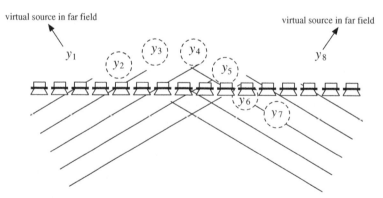

Figure 10.13   Each signal corresponding to the front sound stage is defined as a virtual source. The independent lateral sound is emitted as plane waves (virtual sources in the far field).

Figure 10.13 for $I = 8$. For each other signal, a virtual source is defined with a location as desired. In the example shown in Figure 10.13, the distance is varied for the different sources and some of the sources are defined to be in the front of the sound emitting array, i.e. the virtual sound stage can be defined with an individual distance for each defined direction.

10.4.5 Modifying the decomposed audio signals

Controlling the width of the sound stage

By modifying the estimated scale factors, e.g. A_b, one can control the width of the virtual sound stage. By linear scaling with a factor larger than one, the instruments being part of the sound stage are moved more to the side. The opposite can be achieved by scaling with a factor smaller than one. Alternatively, one can modify the amplitude panning law Equation (10.20) for computing the angle of the direct localized sound.

Modifying the ratio between direct localized sound and the independent sound

For controlling the amount of ambience one can scale the independent lateral sound signals $\hat{n}_1'$ and $\hat{n}_2'$ for getting more or less ambience. Similarly, the localized direct sound can be modified in strength by means of scaling the $\hat{s}'$ signals.

10.5 Subjective evaluation

Informal listening indicated that the scheme described here offers a benefit over conventional stereo playback, especially for off-sweet-spot listening. The goal for the subjective test was not to gain specific psychophysically interesting data, but to get some evidence that the scheme is preferred by listeners compared with conventional stereo playback.

10.5.1 Subjects and playback setup

Eight subjects participated in the tests. Six of these subjects had already participated in the past in subjective tests for audio quality evaluation. The subjects all had an age between 26 and 37 years and reported normal hearing. The test was carried out in a sound insulated room mimicking a typical living room. For audio playback a laptop computer (Apple PowerBook G4) was used with an external D/A converter (MOTU 896) connected directly to eight active loudspeakers (Genelec 1029A).

The loudspeakers were arranged in front of the subject, as illustrated in Figure 10.14. All loudspeakers were always switched on and the subjects had no explicit knowledge from which loudspeakers sound was emitted.

10.5.2 Stimuli

Eleven different stereo music clips were selected. The clips were obtained from CDs and were of lengths between 10 and 15. In order to demonstrate that the described scheme performs well for audio material which is encoded with a typical audio coder, the clips were encoded using MP3 [40] at 192 kb/s. The MP3 encoder integrated with Apple QuickTime 6 was used.

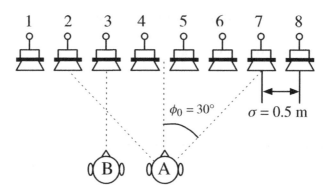

Figure 10.14   A subjective test was carried out with a loudspeaker setup as shown.

Three of the clips were used as training items and the other eight clips were used in the test. The clips contained classical, jazz, rock/pop, and latin music.

Each of the clips was processed to generate two types of eight-channel stimuli. One stimulus type, denoted standard stereo (SS), emits the stereo signal out of loudspeakers 2 and 7, mimicking a standard stereo configuration. The second type of stimulus, denoted front array (FA), is processed according to Equation (10.27) such that the virtual sound stage is reproduced with loudspeakers 2–7 and plane waves with angles $\pm 40°$ are reproduced with loudspeakers 1–8.

10.5.3 Test method

Each subject conducted the test twice, directly after each other, with a different listening position. The two listening positions are indicated as A and B in Figure 10.14. In position A, the subject was located centered such that loudspeakers 2–7 formed a standard stereo listening setup with $\phi_0 = 30°$. In position B, the subject was more to the side, i.e. at the lateral position of loudspeaker 3. It was indicated to the listeners that the virtual stage ranges from loudspeakers 2–7.

The subjects were asked to grade different specific properties and the overall audio quality of the processed clips. For each corresponding stimulus pair, SS and FA, one stimulus had to be graded relative to the other one (reference), where either SS or FA was with 50% chance declared to be the reference. Randomization in term of declaring SS or FA as the reference and the ordering of the clips was carried out for each subject individually. The three different grading tasks of the test are summarized in Table 10.1. Task 1 assesses the quality of the virtual sound stage. Task 2 evaluates distortions introduced by the processing that are not related to the spatial aspect of sound. Task 3 assesses the overall audio quality. Note that for all three tasks the ITU-R 7-grade comparison scale [149], shown in Table 10.2, was used.

Before the test, the subject was given written instructions. Then, a short training session with three clips was carried out, followed by the two tests (listener in position A and B) containing the eight clips listed in Table 10.3.

Table 10.1 Tasks and scales of the subjective test.

Task		Scale
1.	Image quality	ITU-R 7-grade comparison
2.	Audio quality (ignoring image quality)	ITU-R 7-grade comparison
3.	Overall quality	ITU-R 7-grade comparison

Table 10.2 The ITU-R 7-grade comparison scale for comparing an item A with a reference item R.

3	A much better R
2	A better R
1	A slightly better R
0	A same R
−1	A slightly worse R
−2	A worse R
−3	A much worse R

Table 10.3 The eight music clips used for the test.

	Name	Genre
a	I will survive	Pop/rock
b	Blue eyes	Pop/rock
c	Bovio	Classical
d	Slavonic dance	Classical
e	Piensa	Latin/jazz
f	He perdido contigo	Latin/jazz
g	Scoppin	Jazz
h	Y tal vez	latin/Jazz

Figure 10.15 illustrates the graphical user interface that was used for the test. The subject was presented with (frozen) sliders for the reference and for the corresponding other stimulus. With the 'Play' buttons the subject could listen to either the reference or corresponding other stimulus. The subject could switch between the stimuli at any time, while the sound instantly faded from one type of stimulus to the other. Informal listening indicated that such instant switching greatly facilitates comparison of the spatial attributes of the stimuli.

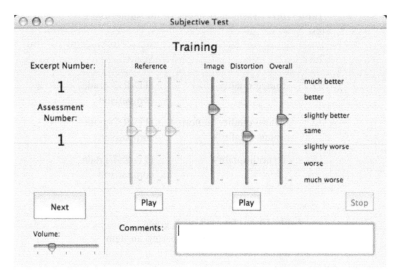

Figure 10.15 The graphical user interface used for the test. The left three (frozen) sliders correspond to the reference and the right three sliders to the corresponding other stimulus.

The duration of the test session (test in Position A, test in Position B) varied between the listeners due to the freedom to repeat the stimuli as often as requested. Typically the test duration was between 30 and 50 minutes.

10.5.4 Results

Figure 10.16 shows the results of the tests with the subjects located at listening position A (sweet spot). The letters indicated on the x-axis correspond to the specific clip

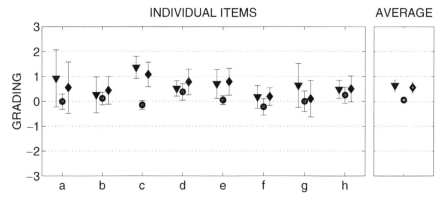

Figure 10.16 The subjective test results for the subjects in position A (sweet spot). The grading and 95% confidence intervals for each clip averaged over all subjects (left) and overall average gradings (right) are shown (triangle = image quality, circle = distortion, diamond = overall quality). Positive gradings indicate that FA is better than SS.

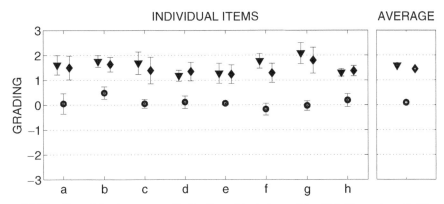

Figure 10.17   The subjective test results for the subjects in position B (off-sweet-spot). The grading and 95% confidence intervals for each clip averaged over all subjects (left) and overall average gradings (right) are shown (triangle = image quality, circle = distortion, diamond = overall quality). Positive gradings indicate that FA is better than SS.

labels given in Table 10.3. The grading scale on the y-axis corresponds to the comparison scale given in Table 10.1, where positive gradings indicate that FA (described scheme) is better than SS (standard stereo). The left panel shows the gradings and 95% confidence intervals for each clip, averaged for all subjects. The right panel shows the results averaged for all clips and subjects. The gradings are shown for the attributes image quality, distortion, and overall quality. The image quality indicates that the subjects preferred the virtual sound stage of the described scheme. The distortion, in most cases close to zero, indicates that the scheme described introduces relatively few distortions if at all. The subjects preferred the described scheme, as is implied by the positive overall quality gradings.

The results for the test with the subjects in the off-sweet-spot position, B, are shown in Figure 10.17. The conclusions here are similar, only that the degree of improvement compared with stereo is significantly larger, as expected, since the virtual stage for stereo with a listener not in the sweet spot is degraded.

10.6 Conclusions

The goal of the subjective test with the subject in position A (sweet spot) was to assess whether the described playback of stereo signals over multiple loudspeakers matches the quality of stereo playback when the listener is located in the sweet spot. On average the subjects preferred the described playback scheme over stereo in the sweet spot.

To show the benefit of listening when the listener is not located in the sweet spot, the subjective test with the listening position B was carried out. As expected, the relative performance of the described scheme is better for off-sweet-spot listening since it maintains the extent of the virtual sound stage.

Only informally tested, the described playback scheme results in a virtual sound stage which hardly depends on the listener's position. The listener can move and the stage and instruments remain at their (absolute) spatial position. This is in contrast to wavefield synthesis systems when they are used for stereo playback. Usually, the left and right stereo signals are emitted as plane waves in such systems, resulting in that the virtual stage moves as the listener moves. The scheme described gives the flexibility of playing back stereo signals over wavefield synthesis systems where the distance of each virtual source (signal) can be freely determined. Previously, a large sweet spot could only be obtained by emitting plane waves, i.e. by mimicking a sound stage infinitely far away.

10.7 Acknowledgement

Sections of text and associated figures in this chapter have been reproduced from C. Faller, "Multi-Loudspeaker Playback of Stereo Signals", J. Audio Engineering Society, November 2006 by permission of the Audio Engineering Society, Inc, New York, USA.

Frequently Used Terms, Abbreviations and Notation

Terms and abbreviations

Anechoic chamber: Type of room with almost totally sound-absorbing walls, frequently used for experimentation under free-field-like conditions.

Auditory object: Perception corresponding to a single sound source. Attributes of auditory objects are location and extent.

Auditory spatial image: Illusion of perception of a space with auditory objects of specific extent and at specific locations.

BCC: Coding of stereo or multi-channel signals using a down-mix and binaural cues (or spatial parameters) to describe inter-channel relationships.

Binaural cues: Inter-channel cues between the left and right ear entrance signals (see also ITD, ILD, and IC).

BMLD: Binaural masking level difference. The difference in masked threshold due to different spatial cues of masker and target signal.

BRIR: Binaural room impulse response, modeling transduction of sound from a source to left and right ear entrances in enclosed spaces. The difference from HTRFs is that BRIRs also consider reflections.

CPC: Channel prediction coefficient.

DFT: Discrete Fourier transform.

Direct sound: Sound reaching a listener's ears or microphones through a direct (nonreflected) path.

Enclosed space: A space where sound is reflected from the walls enclosing it.

Spatial Audio Processing: MPEG Surround and Other Applications   Jeroen Breebaart and Christof Faller
© 2007 John Wiley & Sons, Ltd

ERB: Equivalent rectangular bandwidth. Bandwidth of the auditory filters estimated from masking experiments.

Externalization: When typical stereo music signals are played back over headphones, the extent of the auditory spatial image is limited to about the size of the head. When playing back headphone signals which realistically mimic ear entrance signals during natural listening, the extent of the auditory spatial image can be very realistic and as large in extent as any auditory spatial image in natural listening. The experience of perceiving an auditory spatial image significantly larger than the head during headphone playback is denoted externalization.

FFT: Fast implementation of the DFT, denoted fast Fourier transform (FFT).

Free-field: An open space with no physical objects from which sound is reflected.

Free-field cues: Binaural cues (ITD and ILD) which occur in a one-source free-field listening scenario.

HRTF: Head-related transfer function, modeling transduction of sound from a source to left and right ear entrances in free-field.

IC: Interaural coherence, i.e. degree of similarity between left and right ear entrance signals. This is sometimes also referred to as IAC or interaural cross-correlation (IACC).

ICC: Inter-channel coherence. Same as IC, but defined more generally between any signal pair (e.g. loudspeaker signal pair, ear entrance signal pair, etc.).

ICPD: Inter-channel phase difference. Average phase difference between a signal pair.

ICLD: Inter-channel level difference. Same as ILD, but defined more generally between any signal pair (e.g. loudspeaker signal pair, ear entrance signal pair, etc.).

ICTD: Inter-channel time difference. Same as ITD, but defined more generally between any signal pair (e.g. loudspeaker signal pair, ear entrance signal pair, etc.).

ILD: Interaural level difference, i.e. level difference between left and right ear entrance signals. This is sometimes also referred to as interaural intensity difference (IID).

IPD: Interaural phase difference, i.e. phase difference between the left and right ear entrance signals.

ITD: Interaural time difference, i.e. time difference between left and right ear entrance signals. This is sometimes also referred to as interaural time delay.

kb/s: Unit for bitrate, kilo-bit per second. Also denoted as kbps.

Lateral: From the side, e.g. lateral reflections are reflections arriving at a listener's ears from the sides.

Lateralization: For headphone playback, a subject's task is usually restricted to identifying the lateral displacement of the projection of the auditory object to the straight line connecting the ear entrances. The relationship between the lateral displacement of the auditory object and attributes of the ear entrance signals is denoted lateralization.

LFE channel: Low-frequency effects channel. Multi-channel surround systems often feature one or more LFE channels for low-frequency sound effects requiring higher sound pressure than can be reproduced by the loudspeakers for the regular audio channels. In movie soundtracks, an LFE channel may for example contain low-frequency parts of explosion sounds.

Listener envelopment: A listener's auditory sense of 'envelopment' or 'spaciousness of the environment'.

Localization: The relation between the location of an auditory object and one or more attributes of a sound event. For example, localization may describe the relation between the direction of a sound source and the direction of the corresponding auditory object.

Mixing: Given a number of source signals (e.g. separately recorded instruments, multi-track recording), the process of generating stereo or multi-channel audio signals intended for spatial audio playback is denoted mixing.

OCPD: Overall channel phase difference. A common phase modification of two or more audio channels.

PDF: Probability density function.

Precedence effect: A number of phenomena related to the auditory system's ability to resolve the direction of a source in the presence of one or more reflections by favoring the 'first wave front' over successively arriving reflections.

QMF: Quadrature mirror filter; specific filter as used in audio coders.

Reflection: Sound that arrives at a listener's ears or microphones indirectly after being reflected one or more times from the surface of physical objects.

Reverberation: The persistence of sound in an enclosed space as a result of multiple reflections.

Reverberation time: Defined as the time it takes for sound energy to decay by 60 dB. The more reverberant a room is, the larger is its reverberation time. Reverberation time is often denoted RT or RT60.

SAOC: Spatial audio object coding.

Sound source: A physical object emitting sound.

Spatial audio: Audio signals which, when played back through an appropriate playback system, evoke an auditory spatial image.

Spatial impression: The impression a listener spontaneously gets about type, size, and other properties of an actual or simulated space.

Spatial cues: Cues relevant for spatial perception. This term is used for cues between pairs of channels of a stereo or multi-channel audio signal (see also ICTD, ICLD, and ICC). Also denoted as spatial parameters or binaural cues.

STFT: Short-time (discrete) Fourier transform.

Sweet spot: Optimal listening position for a stereo or multi-channel loudspeaker-based audio playback system.

Transparent: An audio signal is transparent when a listener can not distinguish between this signal and a reference signal. For example, transparent audio coding denotes audio coding, where there is no perceptible degradation in the coded audio signals.

Notation and variables

$\star$	Convolution operator
b	Partition (parameter band) index
b_m	Parameter band b corresponding to sub-band signal m
f_s	Sampling frequency
T	Time constant of a one-sided exponential estimation window
k	Time index of subband signals
	(also time index of STFT spectra)
C	Number of encoder input channels
D	Number of decoder output channels
	(if different from number of encoder input channels C)
E	Number of transmitted channels
	(if different from number of encoder input channels C)
$x_c(n)$	Encoder input audio channels
$s(n)$	Transmitted sum signal or down-mix signal
$s_i(n)$	Down-mix signal i
$y_c(n)$	Transmitted audio channels
$d_i(n)$	Residual signal
$e_i(n)$	Externally provided down-mix signal
$\hat{x}_i(n)$	Decoder output audio channels

$\tilde{x}_i(k)$	One sub-band signal of $x_i(n)$
	(similarly defined for other signals)
$x_{i,m}(k)$	mth sub-band signal of $x_i(n)$
$\hat{x}_{i,m}(k)$	Estimated mth sub-band signal of $x_{i,m}(k)$
$D(s_{i,m})(k)$	Decorrelated version of mth sub-band signal of $s_{i,m}(k)$
$p_{x_i}(k)$	Short-time estimate of power of $x_i(k)$ (similarly defined for other signals)
$p_{x_i,b}$	Frame-based estimate of power of $x_i(k)$ in sub-band b
$h_{l,i}(t)$	Left-ear HRIR for sound source i
$H_{l,i}(f)$	Left-ear HRTF for sound source i
$H_{r,i}(f)$	Right-ear HRTF for sound source i
$\tau_{i_1 i_2}(k)$	ICTD between channel i_1 and i_2
$\Delta L_{i_1 i_2}(k)$	ICLD between channel i_1 and i_2
$c_{i_1 i_2}(k)$	Coherence between channel i_1 and i_2
$\rho_{i_1 i_2}(k)$	Correlation between channel i_1 and i_2
$\Delta L_{i_1 i_2,b}(k)$	ICLD between channel i_1 and i_2 of parameter band b
$\phi_{i_1 i_2,b}(k)$	ICPD between channel i_1 and i_2 of parameter band b
$c_{i_1 i_2,b}(k)$	ICC between channel i_1 and i_2 in parameter band b
$\theta_{i_1 i2,b}(k)$	OCPD between channel i_1 and i_2 of parameter band b
$\tau(k)$	ITD in specific critical band
$\tau_b(k)$	ITD in band parameter band b
$\Delta L(k)$	ILD in specific critical band
$\Delta L_b(k)$	ILD in parameter band b
$c(k)$	IC in specific critical band
$c_b(k)$	IC in parameter band b
	(similarly defined for other signals)
$\gamma_{i,b}$	ith CPC in parameter band b
P_i	Parameter set P of element i

Bibliography

[1] 3rd Generation Partnership Project. 3GPP TS 26.401 V6.2.0, 3rd Generation Partnership Project; Technical Specification Group Services and System Aspects; General audio codec audio processing functions; Enhanced aacPlus general audio codec; General description, March 2005.

[2] V. Algazi, R. Duda, and D. Thompson. The cipic hrtf database, 2001. URL citeseer.ist.psu.edu/article/algazi01cipic.html.

[3] Y. Ando and Y. Kurihara. Nonlinear response in evaluating the subjective diffuseness of sound fields. *J. Acoust. Soc. Am.*, 80(3):833–836, September 1986.

[4] B. S. Atal and S. L. Hanauer. Speech analysis and synthesis by linear prediction of the speech wave. *J. Acoust. Soc. Am.*, 50:637–655, 1971.

[5] B. S. Atal and M. R. Schroeder. Nachahmung der Raumakustik durch Elektronenrechner (Simulation of room acoustics using electronic computers). *Gravesaner Bl"atter*, 27/28: 125–137, 1966.

[6] M. Barron and A. H. Marshall. Spatial impression due to early lateral reflections in concert halls: the derivation of a physical measure. *J. of Sound and Vibration*, 77(2):211–232, 1981.

[7] R. Batra, S. Kuwada, and D. C. Fitzpatrick. Sensitivity to interaural temporal disparities of low- and high-frequency neurons in the superior olivary complex. I. Heterogeneity of responses. *J. Neurophysiol.*, 78:1222–1236, 1997.

[8] R. Batra, S. Kuwada, and D. C. Fitzpatrick. Sensitivity to interaural temporal disparities of low- and high-frequency neurons in the superior olivary complex. II. Coincidence detection. *J. Neurophysiol.*, 78:1237–1247, 1997.

[9] J. Bauck. Conversion of two-channel stereo for presentation by three frontal loudspeakers. In: *Preprint 109th Conv. Aud. Eng. Soc.*, Sept. 2000.

[10] B. B. Bauer. Stereophonic earphones and binaural loudspeakers. *J. Audio Eng. Soc.*, 9: 148–151, 1961.

[11] B. B. Bauer. Phasor analysis of some stereophonic phenomena. *J. Acoust. Soc. Am.*, 33: 1536–1539, Nov. 1961.

[12] A. Baumgarte, C. Faller, and P. Kroon. Audio coder enhancement using scalable binaural cue coding with equalized mixing. In: *Preprint 116th Conv. Aud. Eng. Soc.*, May 2004.

[13] F. Baumgarte and C. Faller. Why Binaural Cue Coding is better than Intensity Stereo Coding. In: *Preprint 112th Conv. Aud. Eng. Soc.*, May 2002.

[14] D. R. Begault. *3-D Sound for Virtual Reality and Multimedia*. Academic Press, Cambridge, MA, 1994.

[15] J. C. Bennett, K. Barker, and F. O. Edeko. A new approach to the assessment of stereophonic sound system performance. *J. Audio Eng. Soc.*, 33(5):314–321, May 1985.

[16] A. J. Berkhout, D. de Vries, and P. Vogel. Wave front synthesis: a new direction in electroacoustics. In: *Preprint 93th Conv. Aud. Eng. Soc.*, Oct. 1992.

[17] A. J. Berkhout, D. de Vries, and P. Vogel. Acoustic control by wave field synthesis. *J. Acoust. Soc. Am.*, 93(5):2764–2778, May 1993.

[18] B. Bernfeld. Attempts for better understanding of the directional stereophonic listening mechanism. In: *Preprint 44th Conv. Aud. Eng. Soc.*, Feb. 1973.

[19] L. R. Bernstein and C. Trahiotis. Discrimination of interaural envelope correlation and its relation to binaural unmasking at high frequencies. *J. Acoust. Soc. Am.*, 91:306–316, 1992.

[20] L. R. Bernstein and C. Trahiotis. The normalized correlation: accounting for binaural detection across center frequency. *J. Acoust. Soc. Am.*, 100:3774–3787, 1996.

[21] L. R. Bernstein and C. Trahiotis. The effects of randomizing values of interaural disparities on discrimination of interaural correlation. *J. Acoust. Soc. Am.*, 102:1113–1119, 1997.

[22] L. R. Bernstein and C. Trahiotis. The effects of signal duration on NoSo and NoSπ thresholds at 500 Hz and 4 kHz. *J. Acoust. Soc. Am.*, 105:1776–1783, 1999.

[23] L. R. Bernstein, S. van de Par, and C. Trahiotis. The normalized interaural correlation: Accounting for NoSπ thresholds obtained with Gaussian and 'low-noise' masking noise. *J. Acoust. Soc. Am.*, 106(2):870–876, August 1999.

[24] F. A. Bilsen and J. Raatgever. Spectral dominance in binaural hearing. *Acustica*, 28:131–132, 1973.

[25] F. A. Bilsen and J. Raatgever. Spectral dominance in binaural lateralization. *Acustica*, 28: 131–132, 1977.

[26] J. Blauert. *Spatial hearing: the psychophysics of human sound localization.* MIT Press, Cambridge, Massachusetts, 1997.

[27] M. A. Blommer and G. H. Wakefield. Pole-zero approximations for HRTF using a logarithmic error criterium. *IEEE Trans. Speech Audio Processing*, 5:278–287, 1997.

[28] A. Blumlein. Improvements in and relating to sound transmission, sound recording and sound reproduction systems. *British Patent Specification 394325*, 1931. Reprinted in *Stereophonic Techniques*, Aud. Eng. Soc., New York, 1986.

[29] S. E. Boehnke, S. E. Hall, and T. Marquardt. Detection of static and dynamic changes in interaural correlation. *J. Acoust. Soc. Am.*, 112:1617–1626, 2002.

[30] G. Boerger, P. Laws, and J. Blauert. Stereophone Kopfhörerwiedergabe mit Steuerung bestimmter Übertragungsfaktoren durch Kopfdrehbewegung (Sereophonic headphone reproduction with variation of various transfer factors by means of rotational head movements). *Acustica*, 39:22–26, 1977.

[31] S. Boland and M. Deriche. High quality audio coding using multipulse LPC and wavelet decomposition. In: *Proc. Int. Conf. Acoust., Speech, and Signal Proc.*, pages 3067–3069, Detroit, USA, May 1995. IEEE.

[32] M. Bosi, K. Brandenburg, S. Quackenbush, L. Fielder, K. Akagiri, H. Fuchs, M. Dietz, J. Herre, G. Davidson, and Y. Oikawa. ISO/IEC MPEG-2 advanced audio coding. *J. Audio Eng. Soc.*, 45(10):789–814, 1997.

[33] J. C. Boudreau and C. Tsuchitani. Binaural interaction in the cat superior olive S segment. *J. Neurophysiol.*, 31:442–454, 1968.

[34] M. Boufi and C. Kyirakakis. Audio signal de-correlation based on a critical band approach. In: *Preprint 117th Conv. Aud. Eng. Soc.*, page preprint 6291, Oct. 2004.

[35] J. Braasch. Localization in the presence of a distracter and reverberation in the frontal horizontal plane. II. Model algorithms. *Acta Acustica United with Acustica*, 88(6):956–969, November/December 2002.

[36] J. Braasch and K. Hartung. Localization in the presence of a distracter and reverberation in the frontal horizontal plane. I. Psychoacoustical data. *Acta Acustica United with Acustica*, 88 (6):942–955, November/December 2002.

[37] J. Braasch, J. Blauert, and T. Djelani. The precedence effect for noise bursts of different bandwidths. I. Psychoacoustical data. *Acoust. Sci. and Tech.*, 24(5):233–241, 2003.

[38] J. S. Bradley. Comparison of concert hall measurements of spatial impression. *J. Acoust. Soc. Am.*, 96(6):3525–3535, 1994.

[39] J. S. Bradley and B. A. Soulodre. Objective measures of listener envelopment. *J. Acoust. Soc. Am.*, 98:2590–2597, 1995.

[40] K. Brandenburg and G. Stoll. ISO-MPEG-1 audio: a generic standard for coding of high-quality digital audio. *J. Audio Eng. Soc.*, pages 780–792, Oct. 1994.

[41] K. Brandenburg, G. G. Langenbucher, H. Schramm, and D. Seitzer. A digital signal processor for real time adaptive transform coding of audio signal up to 20 khz bandwidth. *Proc. ICCC*, pages 474–477, 1982.

[42] D. J. Breebaart. *Modeling binaural signal detection.* PhD thesis, Eindhoven University of Technology, Eindhoven, 2001.

[43] J. Breebaart and A. Kohlrausch. The Perceptual (ir)relevance of HRTF magnitude and phase spectra. In: *Preprint 5406, 110th AES convention*, Amsterdam, The Netherlands, 2001.

[44] J. Breebaart, S. van de Par, and A. Kohlrausch. Binaural signal detection with phase-shifted and time-delayed noise maskers. *J. Acoust. Soc. Am.*, 103:2079–2083, 1998.

[45] J. Breebaart, S. van de Par, and A. Kohlrausch. The contribution of static and dynamically varying ITDs and IIDs to binaural detection. *J. Acoust. Soc. Am.*, 106:979–992, 1999.

[46] J. Breebaart, S. van de Par, and A. Kohlrausch. Binaural processing model based on contralateral inhibition. I. Model setup. *J. Acoust. Soc. Am.*, 110:1074–1088, 2001.

[47] J. Breebaart, S. van de Par, A. Kohlrausch, and E. Schuijers. Parametric coding of stereo audio. *EURASIP J. on Applied Signal Processing*, 9:1305–1322, 2004.

[48] J. Breebaart, J. Herre, C. Faller, J. Röden, F. Myburg, S. Disch, H. Purnhagen, G. Hotho, M. Neusinger, K. Kjörling, and W. Oomen. MPEG spatial audio coding / MPEG Surround: Overview and current status. In: *119th AES convention*, New York, USA, 2005.

[49] J. Breebaart, J. Herre, L. Villemoes, C. Jin, K. Kjörling, and J. Plogsties. Multichannel goes mobile: MPEG Surround binaural rendering. In: *Proc. 29th AES conference*, Seoul, Korea, 2006.

[50] A. Bronkhorst and T. Houtgast. Auditory distance perception in rooms. *Nature*, 397:517–520, Feb. 1999.

[51] A. A. M. L. Bruekers, W. J. Oomen, and R. J. van der Vleuten. Lossless coding for DVD audio. In: *Preprint 100th Conv. Aud. Eng. Soc.*, November 1996.

[52] D. S. Brungart and W. M. Rabinowitz. Auditory localization of nearby sources I: Head-related transfer functions. *J. Acoust. Soc. Am.*, 106:1465–1479, 1999.

[53] D. S. Brungart, N. I. Durlach, and W. M. Rabinowitz. Auditory localization of nearby sources. II. Localization of a broadband source. *J. Acoust. Soc. Am.*, 106:1956–1968, 1999.

[54] S. Carlile, C. Jin, and V. Van Raad. Continuous virtual auditory space using HRTF interpolation: acoustic and phychophysical errors. In: *Int. Symposium on multimedia information processing*, Sydney, Australia, 2000.

[55] C. Cellier, P. Chênes, and M. Rossi. Lossless audio data compression for real time applications. In: *Preprint 95th Conv. Aud. Eng. Soc.*, Oct. 1993.

[56] J. Chen, B. D. Van Veen, and K. E. Hecox. A spatial feature extraction and regularization model for the head-related transfer function. *J. Acoust. Soc. Am.*, 97:439–452, 1995.

[57] J.-H. Chen and D. Wang. Transform predictive coding of wideband speech signals. In: *Proc. ICASSP*, pages 275–278, May 1996.

[58] R. I. Chernyak and N. A. Dubrovsky. Pattern of the noise images and the binaural summation of loudness for the different interaural correlation of noise. In: *Proc. 6th Int. Congr. on Acoustics Tokyo*, volume 1, pages A–3–12, 1968.

[59] N. Cheung, S. Trautmann, and A. Horner. Head-related transfer function modeling in 3-D sound systems with genetic algorithms. *J. Audio Eng. Soc.*, 46:531–539, 1998.

[60] H. S. Colburn. Theory of binaural interaction based on auditory-nerve data. II. Detection of tones in noise. *J. Acoust. Soc. Am.*, 61:525–533, 1977.

[61] H. S. Colburn and N. I. Durlach. Models of binaural interaction. In: E.C. Carterette and M.P. Friedman (Ed), *Handbook of Perception*, volume IV: Hearing, pages 467–518. Academic Press, New York, San Francisco, London, 1978.

[62] P. D. Coleman. Failure to localize the source distance of an unfamiliar sound. *J. Acoust. Soc. Am.*, 34:345–346, 1962.

[63] J. F. Culling, H. S. Colburn, and M. Spurchise. Interaural correlation sensitivity. *J. Acoust. Soc. Am.*, 110:1020–1029, 2001.

[64] P. Damaske. Subjektive Untersuchungen von Schallfeldern (Subjective investigations of sound fields). *Acustica*, 19:198–213, 1967/68.

[65] P. Damaske. Die psychologische Auswertung akustischer Phänomeme (The psychological interpretation of acoustical phenomena). In: *Proc. 7th Int. Congr. on Acoustics*, volume 21 G 2, Budapest, 1971. 7th Int. Congr. on Acoustics.

[66] A.C. den Brinker, E.G.P. Schuijers, and A.W.J. Oomen. Parametric coding for high-quality audio. In: *AES 112th Convention*, Munich, DE, May 2002.

[67] B. den Brinker, E. Schuijers, and W. Oomen. Parametric coding for high-quality audio. In: *Preprint 112th Conv. Aud. Eng. Soc.*, May 2002.

[68] M. Dietz, L. Liljeryd, K. Kjörling, and O. Kunz. Spectral band replication, a novel approach in audio coding. In: *Preprint 5553, 112th AES Convention, Munich*, 2002.

[69] R. Dressler. Dolby Surround Prologic II Decoder – Principles of operation. Technical report, Dolby Laboratories, 2000. www.dolby.com/tech/.

[70] R. Drullman and A. W. Bronkhorst. Multichannel speech intelligibility and talker recognition using monaural, binaural, and three-dimensional auditory presentation. *J. Acoust. Soc. Am.*, 107(4):2224–2235, April 2000.

[71] N. I. Durlach. Equalization and cancellation theory of binaural masking-level differences. *J. Acoust. Soc. Am.*, 35:1206–1218, 1963.

[72] N. I. Durlach and H. S. Colburn. Binaural phenomena. In: E.C. Carterette and M.P. Friedman, editors, *Handbook of Perception*, volume IV: Hearing, pages 365–466. Academic Press, New York, San Francisco, London, 1978.

[73] N. I. Durlach and X. D. Pang. Interaural magnification. *J. Acoust. Soc. Am.*, 80:1849–1850, 1986.

[74] J. Eargle, editor. *Stereophonic Techniques*. Audio Engineering Society, New York, 1986.

[75] J. M. Eargle. Multichannel stereo matrix systems: An overview. *IEEE Trans. on Speech and Audio Proc.*, 19(7):552–559, July 1971.

[76] B. Edler, H. Purnhagen, and C. Ferekidis. An analysis/synthesis audio codec (asac). In: *Preprint 100th Conv. Aud. Eng. Soc.*, May 1996.

[77] A. Ehret, K. Kjörling, J. Röden, H. Purnhagen, and H. Hörich. aacPlus, only a low-bitrate codec? In: *AES 117th Convention*, San Francisco, US, October 2004.

[78] P. Ekstrand. Bandwidth extension of audio signals by spectral band replication. In: *Proc. 1st IEEE Benelux Workshop on Model based Processing and Coding of Audio (MPCA-2002)*, Leuven, BE, November 2002.

[79] J. Engdegård, H. Purnhagen, J. Rödén, and L. Liljeryd. Synthetic ambience in parametric stereo coding. In: *Proc. 116th AES convention, Berlin, Germany*, 2004.

[80] F. Enkl. Die übertragung räumlicher Schallfeldstrukturen über einen kanal mit hilfe unter-schwelliger pilotfrequenzen (The transmission of spatial sound field structures over one channel aided by pilot tones below the threshold). *Elektron. Rdsch.*, 12:347–349, 1958.

[81] M. J. Evans, J. A. S. Angus, and A. I. Tew. Analyzing head-related transfer function mea-surements using surface spherical harmonics. *J. Acoust. Soc. Am.*, 104:2400–2411, 1998.

[82] C. Faller. Parametric multi-channel audio coding: Synthesis of coherence cues. *IEEE Trans. on Speech and Audio Proc.*, 14(1):299–310, Jan. 2006.

[83] C. Faller. *Parametric Coding of Spatial Audio*. PhD thesis, Ecole Polytechnique Fédérale de Lausanne (EPFL), Switzerland, July 2004. Thesis No. 3062, http://library.epfl.ch/theses/ ?nr=3062.

[84] C. Faller. Parametric joint-coding of audio sources. In: *Preprint 120th Conv. Aud. Eng. Soc.*, May 2006.

[85] C. Faller. Multi-loudspeaker playback of stereo signals. *J. of the Aud. Eng. Soc.*, 54(11): 1051–1064, Nov. 2006.

[86] C. Faller and F. Baumgarte. Binaural Cue Coding – Part II: Schemes and applications. *IEEE Trans. on Speech and Audio Proc.*, 11(6):520–531, Nov. 2003.

[87] C. Faller and J. Merimaa. Source localization in complex listening situations: Selection of binaural cues based on interaural coherence. *J. Acoust. Soc. Am.*, 116(5):3075–3089, Nov. 2004.

[88] L. D. Fielder, M. Bosi, G. Davidson, M. Davis, C. Todd, and S. Vernon. AC-2 and AC-3: Low-complexity transform-based audio coding. In: N. Gilchrist and C. Grewin, editors, *Collected Papers on Digital Audio Bit-Rate Reduction*, pages 54–72. Audio Engineering Society Inc., 1996.

[89] K. J. Gabriel and H. S. Colburn. Interaural correlation discrimination: I. Bandwidth and level dependence. *J. Acoust. Soc. Am.*, 69:1394–1401, 1981.

[90] W. Gaik. Combined evaluation of interaural time and intensity differences: psychoacoustic results and computer modeling. *J. Acoust. Soc. Am.*, 94:98–110, 1993.

[91] B. Gardner and K. Martin. HRTF Measurements of a KEMAR dummy-head microphone. MIT Media Lab Perceptual Computing Technical Report 280, May 1994, 1994. URL cite-seer.ist.psu.edu/gardner94hrtf.html.

[92] R. Geiger, T. Sporer, J. Koller, and K. Brandenburg. Audio coding based on integer trans-form. In: *Preprint 111th Conv. Aud. Eng. Soc.*, Nov. 2001.

[93] M. Gerzon. Three channels: the future of stereo? *Studio Sound*, pages 112–125, June 1990.

[94] M. A. Gerzon. Optimal reproduction matrices for multispeaker stereo. In: *Preprint 91st Conv. Aud. Eng. Soc.*, Oct. 1991.

[95] M. A. Gerzon, P. G. Graven, J. R. Stuart, M. J. Law, and R. J. Wilson. The MLP lossless compression system. In: *Proc. AES 17th Int. Conf.: High-Quality Audio Coding*, pages 61–75, Florence, Italy, September 1999. AES.

[96] C. Giguère and S. M. Abel. Sound localization: Effects of reverberation time, speaker array, stimulus frequence, and stimulus rise/decay. *J. Acoust. Soc. Am.*, 94(2):769–776, August 1993.

[97] B. R. Glasberg and B. C. J. Moore. Auditory filter shapes in forward masking as function of level. *J. Acoust. Soc. Am.*, 71:946–949, 1982.

[98] B. R. Glasberg and B. C. J. Moore. Derivation of auditory filter shapes from notched-noise data. *Hearing Research*, 47:103–138, 1990.

[99] M. D. Good and R. H. Gilkey. Sound localization in noise: The effect of signal to noise ratio. *J. Acoust. Soc. Am.*, 99(2):1108–1117, February 1996.

[100] M. D. Good, R. H. Gilkey, and J. M. Ball. The relation between detection in noise and localization in noise in the free field. In Robert H. Gilkey and Timothy R. Anderson, editors, *Binaural and Spatial Hearing in Real and Virtual Environments*, pages 349–376. Lawrence Erlbaum Associates, Mahwah, NJ, USA, 1997.

[101] D. W. Grantham. Interaural intensity discrimination: insensitivity at 1000 Hz. *J. Acoust. Soc. Am.*, 75:1191–1194, 1984.

[102] D. W. Grantham and D. E. Robinson. Role of dynamic cues in monaural and binaural signal detection. *J. Acoust. Soc. Am.*, 61:542–551, 1977.

[103] D. M. Green. Signal-detection analysis of equalization and cancellation model. *J. Acoust. Soc. Am.*, 40:833–838, 1966.

[104] B. Grill. The MPEG-4 general audio coder. In: *Proc. AES 17th Int. Conf.: High-Quality Audio Coding*, pages 147–156, Florence, Italy, September 1999. AES.

[105] A. Gröschel, M. Schug, M. Beer, and F. Henn. Enhancing audio coding efficiency of mpeg layer-2 with spectral band replication for digital radio (dab) in a backwards compatible way. In: *AES 114th Convention*, Amsterdam, NL, March 2003.

[106] K. Gundry. A new active matrix decoder for surround sound. In: *Proc. AES 19th Int. Conf.*, June 2001.

[107] E. R. Hafter and S. C. Carrier. Masking-level differences obtained with pulsed tonal maskers. *J. Acoust. Soc. Am.*, 47:1041–1047, 1970.

[108] J. W. Hall and M. A. Fernandes. The role of monaural frequency selectivity in binaural analysis. *J. Acoust. Soc. Am.*, 76:435–439, 1984.

[109] J. W. Hall and A. D. G. Harvey. NoSo and NoSπ thresholds as a function of masker level for narrow-band and wideband masking noise. *J. Acoust. Soc. Am.*, 76:1699–1703, 1984.

[110] A. Härmä and U. K. Laine. Warped low-delay celp for wide-band audio coding. In: *Proc. of the AES 17th Int. Conference: High-Quality Audio Coding*, pages 207–215, Sept. 1999.

[111] A. Härmä, U. K. Laine, and M. Karjalainen. An experimental audio codec based on warped linear prediction of complex valued signals. In: *Proc. ICASSP 1997*, volume 1, pages 323–327, Apr. 1997.

[112] W. M. Hartmann. Localization of sound in rooms. *J. Acoust. Soc. Am.*, 74(5):1380–1391, Nov. 1983.

[113] W. M. Hartmann. Listening in a room and the precedence effect. In Robert H. Gilkey and Timothy R. Anderson, editors, *Binaural and Spatial Hearing in Real and Virtual Environments*, pages 349–376. Lawrence Erlbaum Associates, Mahwah, NJ, USA, 1997.

[114] W. M. Hartmann and B. Rakerd. Localization of sound in rooms, IV: The Franssen effect. *J. Acoust. Soc. Am.*, 86(4):1366–1373, Oct. 1989.

[115] W. M. Hartmann and A. Wittenberg. On the externalization of sound images. *J. Acoust. Soc. Am.*, 99:3678–3688, 1996.

[116] M. L. Hawley, R. Y. Litovsky, and H. S. Colburn. Speech intelligibility and localization in a multi-source environment. *J. Acoust. Soc. Am.*, 105(5):3436–3448, June 1999.

[117] S. Haykin. *Adaptive Filter Theory (third edition)*. Prentice Hall, 1996.

[118] P. Hedelin. A tone-oriented voice-excited vocoder. In: *Proc. IEEE Int. Conf. Acoust., Speech, and Signal Proc.*, volume I, pages 205–208, Atlanta, USA, March 1981. IEEE.

[119] J. Herre, K. Brandenburg, and D. Lederer. Intensity stereo coding. *96th AES Conv., Feb. 1994, Amsterdam (preprint 3799)*, 1994.

[120] J. Herre, C. Faller, C. Ertel, J. Hilpert, A. Hoelzer, and C. Spenger. MP3 Surround: Efficient and Compatible Coding of Multi-Channel Audio. In: *Proc. 116th AES convention*, Berlin, Germany, 2004.

[121] J. Herre, H. Purnhagen, J. Breebaart, C. Faller, S. Disch, K. Kjörling, E. Schuijers, J. Hilpert, and F. Myburg. The reference model architecture of MPEG spatial audio coding. In: *118th AES convention*, Barcelona, Spain, 2005.

[122] J. Herre, K. Kjörling, J. Breebaart, C. Faller, K. S. Chon, S. Disch, H. Purnhagen, J. Koppens, J. Hilpert, J. Röden, W. Oomen, K. Linzmeier, and L. Villemoes. MPEG Surround – The ISO/MPEG standard for efficient and compatible multi-channel audio coding. In: *Proc. 122th AES convention*, Vienna, Austria, 2007.

[123] R.M. Hershkowitz and N.I. Durlach. Interaural time and amplitude jnds for a 500-hz tone. *J. Acoust. Soc. Am.*, 46:1464–1467, 1969.

[124] I.J. Hirsh. The influence of interaural phase on interaural summation and inhibition. *J. Acoust. Soc. Am.*, 20:536–544, 1948.

[125] I. Holube, M. Kinkel, and B. Kollmeier. Binaural and monaural auditory filter bandwidths and time constants in probe tone detection experiments. *J. Acoust. Soc. Am.*, 104:2412–2425, 1998.

[126] G. Hotho, L. Villemoes, and J. Breebaart. A stereo backward compatible multi-channel audio codec. *IEEE Transactions on Audio, Speech and Language processing*, page submitted, 2007.

[127] A. J. M. Houtsma, C. Trahiotis, R. N. J. Veldhuis, and R. van der Waal. Bit rate reduction and binaural masking release in digital coding of stereo sound. *Acustica / acta acustica*, 92: 908–909, 1996.

[128] A. J. M. Houtsma, C. Trahiotis, R. N. J. Veldhuis, and R. van der Waal. Further bit rate reduction through binaural processing. *Acustica / acta acustica*, 92:909–910, 1996.

[129] J. Hull. Surround sound past, present, and future. Technical report, Dolby Laboratories, 1999. www.dolby.com/tech/.

[130] J. Huopaniemi. *Virtual Acoustics and 3D Sound in Multimedia Signal Processing*. PhD thesis, Laboratory of Acoustics and Audio Signal Processing, Helsinki University of Technology, Finland, 1999. Rep. 53.

[131] J. Huopaniemi and N. Zacharov. Objective and subjective evaluation of head-related transfer function filter design. *J. Audio. Eng. Soc.*, 47:218–239, 1999.

[132] IEEE. IEEE recommended practice for speech quality measurements. *IEEE Trans. Audio Electroacoust.*, 17(3):137–148, 1969.

[133] D. R. F. Irvine and G. Gago. Binaural interaction in high-frequency neurons in inferior colliculus of the cat: effects of variations in sound pressure level on sensitivity to interaural intensity differences. *J. Neurophysiol.*, 63:570–591, 1990.

[134] R. Irwan and R. M. Aarts. Two-to-five channel sound processing. *J. Audio Eng. Soc.*, 50: 914–926, 2002.

[135] ISO/IEC JTC1/SC29/WG11. Coding of audio-visual objects. Part 3: Audio, AMENDMENT 2: Parametric coding of high quality audio. ISO/IEC Int. Std. 14496-3:2001/Amd2:2004, July 2004.

[136] ISO/IEC JTC1/SC29/WG11. Coding of moving pictures and associated audio for digital storage media at up to about 1.5 Mbit/s – Part 3: Audio. ISO/IEC 11172-3 International Standard 5, 1993.

[137] ISO/IEC JTC1/SC29/WG11. Generic coding of moving pictures and associated audio information – Part 7: Advanced Audio Coding. ISO/IEC 13818-7 International Standard, 1997.

[138] ISO/IEC JTC1/SC29/WG11. MPEG audio technologies – Part 1: MPEG Surround. ISO/IEC FDIS 23003-1:2006(E), 2004.

[139] ISO/IEC JTC1/SC29/WG11. MPEG-4 Overview. JTC1/SC29/WG11 N4668, 2002.

[140] ISO/IEC JTC1/SC29/WG11. MPEG-4 Audio Version 2. ISO/IEC 14496-3 International Standard, 1999.

[141] ISO/IEC JTC1/SC29/WG11. Report on the verification tests of MPEG-4 High Efficiency AAC. ISO/IEC JTC1/SC29/WG11 N6009, October 2003.

[142] ISO/IEC JTC1/SC29/WG11. Call for Proposals on Spatial Audio Coding. ISO/IEC JTC1/SC29/WG11 N6455, October 2004.

[143] ISO/IEC JTC1/SC29/WG11. Report on the verification test of MPEG-4 parametric coding for high-quality audio. ISO/IEC JTC1/SC29/WG11 N6675, 2004.

[144] ISO/IEC JTC1/SC29/WG11. Report on the Verification Tests of MPEG-D MPEG Surround. ISO/IEC JTC1/SC29/WG11 N8851, January 2007.

[145] ISO/IEC JTC1/SC29/WG11. Coding of audio-visual objects – Part 3: Audio, AMENDMENT 1: Bandwidth Extension. ISO/IEC Int. Std. 14496-3:2001/Amd.1:2003, 2003.

[146] ISO/IEC JTC1/SC29/WG11. Subpart 5: MPEG-4 Structured Audio. Final Commitee Draft FCD 14496-3: Coding of Audiovisual Objects, Part 3: Audio, October 1998.

[147] ITU-R. Methods for the subjective assessment of small impairments in audio systems including multichannel sound systems. ITU-R Recommend. BS.1116-1, 1997.

[148] ITU-R. Method for the subjective assessment of intermediate quality level of coding systems (MUSHRA). ITU-R Recommend. BS.1534, 2001.

[149] ITU-R. Subjective assessment of sound quality. ITU-R Recommend. BS.562.3, 1994.

[150] ITU-R. Multichannel stereophonic sound system with and without accompanying picture. ITU-R Recommend. BS.775-1, 1994.

[151] W. H. Janovsky. An apparatus for three-dimensional reproduction in electroacoustical presentations. *German Federal Republic Patent No. 973570*, 1948.

[152] L. A. Jeffress. A place theory of sound localization. *J. Comp. Physiol. Psych.*, 41:35–39, 1948.

[153] L. A. Jeffress and D. McFadden. Differences of interaural phase and level in detection and lateralization. *J. Acoust. Soc. Am.*, 49:1169–1179, 1971.

[154] L. A. Jeffress, H. C. Blodgett, and B. H. Deatherage. Masking and interaural phase. II. 167 cycles. *J. Acoust. Soc. Am.*, 34:1124–1126, 1962.

[155] D. H. Johnson. The relationship between spike rate and synchrony in responses of auditory-nerve fibers to single tones. *J. Acoust. Soc. Am.*, 68:1115–1122, 1980.

[156] J. D. Johnston and A. J. Ferreira. Sum-difference stereo transform coding. In: *Proc. ICASSP-92*, pages 569–572, 1992.

[157] P. X. Joris. Envelope coding in the lateral superior olive. II. Characteristic delays and comparison with responses in the medial superior olive. *J. Neurophysiol.*, 76:2137–2156, 1996.

[158] P. X. Joris and T. C. T. Yin. Envelope coding in the lateral superior olive. I. Sensitivity to interaural time differences. *J. Neurophysiol.*, 73:1043–1062, 1995.

[159] J-M. Jot, M. Walsh, and A. Philp. Binaural simulation of complex acoustic scenes for interactive audio. In: *Convention paper 6950, 121st AES convention*, San Francisco, USA, October 2006.

[160] N. Y. S. Kiang. *The Nervous System*, volume 3, chapter Stimulus representation in the discharge patterns of auditory neurons. Raven Press, New York, 1975.

[161] W. G. King and D. A. Laird. The effect of noise intensity and pattern on locating sounds. *J. Acoust. Soc. Am.*, 2:99–102, 1930.

[162] D. J. Kistler and F. L. Wightman. A model of head-related transfer functions based on principal components analysis and minimum-phase reconstruction. *J. Acoust. Soc. Am.*, 91: 1637–1647, 1992.

[163] W. B. Kleijn and K. K. Paliwal. *An Introduction to Speech Coding*. Elsevier, Amsterdam, 1995.

[164] P. W. Klipsch. Stereophonic sound with two tracks, three channels by means of a phantom circuit (2ph3). *J. Aud. Eng. Soc.*, 6(2):118–123, April 1958.

[165] R. G. Klumpp and H. R. Eady. Some measurements of interaural time difference thresholds. *J. Acoust. Soc. Am.*, 28:859–860, 1956.

[166] A. Kohlrausch. Auditory filter shape derived from binaural masking experiments. *J. Acoust. Soc. Am.*, 84:573–583, 1988.

[167] B. Kollmeier and R. H. Gilkey. Binaural forward and backward masking: evidence for sluggishness in binaural detection. *J. Acoust. Soc. Am.*, 87:1709–1719, 1990.

[168] B. Kollmeier and I Holube. Auditory filter bandwidths in binaural and monaural listening conditions. *J. Acoust. Soc. Am.*, 92:1889–1901, 1992.

[169] S. Komiyama. Subjective evaluation of angular displacement between picture and sound directions for HDTV sound systems. *J. Aud. Eng. Soc.*, 37(4):210–214, April 1989.

[170] A. Kulkarni and H. S. Colburn. Role of spectral detail in sound-source localization. *Nature*, 396:747–749, 1998.

[171] A. Kulkarni, S. K. Isabelle, and H. S. Colburn. Sensitivity of human subjects to head-related transfer-function phase spectra. *J. Acoust. Soc. Am.*, 105:2821–2840, 1999.

[172] O. Kunz. Enhancing MPEG-4 AAC by spectral band replication. In: *Technical sessions proceedings of Workshop and exhibition on MPEG-4 (WEMP4), San Jose Fairmont (USA)*, pages 41–44, 2002.

[173] K. Kurozumi and K. Ohgushi. The relationship between the cross-correlation coefficient of two-channel acoustic signals and sound image quality), and apparent source width (asw) in concert halls. *J. Acoust. Soc. Am.*, 74(6):1726–1733, Dec. 1983.

[174] H. Kuttruff. *Room Acoustics*. Elsevier Science Publishers, Barking, UK, 1991.

[175] S. Kuwada, T. C. T. Yin, J. Syka, T. J. F. Buunen, and R. E. Wickesberg. Binaural interaction in low-frequency neurons in inferior colliculus of the cat. IV. Comparison of monaural and binaural response properties. *J. Neurophysiol.*, 51:1306–1325, 1984.

[176] E. A. H. Langendijk, D. J. Kistler, and F. L. Wightman. Sound localization in the presence of one or two distracters. *J. Acoust. Soc. Am.*, 109(5):2123–2134, May 2001.

[177] E. H. A. Langendijk and A. W. Bronkhorst. Fidelity of three-dimensional-sound reproduction using a virtual auditory display. *J. Acoust. Soc. Am.*, 107:528–537, 2000.

[178] T. L. Langford and L. A. Jeffress. Effect of noise crosscorrelation on binaural signal detection. *J. Acoust. Soc. Am.*, 36:1455–1458, 1964.

[179] H. Lauridsen. Experiments concerning different kinds of room-acoustics recording. *Ingenioren*, 47, 1954.

[180] H. Lauridsen and F. Schlegel. Stereofonie und richtungsdiffuse Klangwiedergabe [Stereophony and directionally diffuse reproduction of sound]. *Gravesaner Blätter*, 5:28–50, 1956.

[181] Y. L. Lee. *Statistical Theory of Communication*. John Wiley, New York, 1960.

[182] W. Lindemann. Extension of a binaural cross-correlation model by contralateral inhibition. I. Simulation of lateralization for stationary signals. *J. Acoust. Soc. Am.*, 80:1608–1622, 1986.

[183] R. Y. Litovsky, H. S. Colburn, W. A. Yost, and S. J. Guzman. The precedence effect. *J. Acoust. Soc. Am.*, 106:1633–1654, 1999.

[184] J. P. A. Lochner and W. de V. Keet. Stereophonic and quasi-stereophonic reproduction. *J. Acoust. Soc. Am.*, 32:392–401, 1960.

[185] C. Lorenzi, S. Gatehouse, and C. Lever. Sound localization in noise in normal-hearing listeners. *J. Acoust. Soc. Am.*, 105(5):1810–1820, March 1999.

[186] J. Makhoul and M. Berouti. High-frequency regeneration in speech coding systems. In: *Proc. ICASSP*, volume 428-431, 1979.

[187] J. C. Makoes and J. C. Middlebrooks. Two-dimensional sound localization by human listeners. *J. Acoust. Soc. Am.*, 87:2188–2200, 1990.

[188] H. S. Malvar. *Signal processing with lapped transforms*. Artech House, 1992.

[189] D. McAlpine, D. Jiang, T. M. Shackleton, and A. R. Palmer. Convergent input from brainstem coincidence detectors onto delay-sensitive neurons in the inferior colliculus. *J. Neurosci.*, 18:6026–6039, 1998.

[190] R. J. McAulay and T. F. Quatieri. Speech analysis/synthesis based on a sinusoidal representation. *IEEE Transactions on Acoustics, Speech, and Signal Processing*, 34(4):744–754, 1986.

[191] D. McFadden, L. A. Jeffress, and H. L. Ermey. Difference in interaural phase and level in detection and lateralization: 250 Hz. *J. Acoust. Soc. Am.*, 50:1484–1493, 1971.

[192] D. H. Mershon and J. N. Bowers. Absolute and relative cues for the auditory perception of egocentric distance. *Perception*, 8:311–322, 1979.

[193] D. H. Mershon and L. E. King. Intensity and reverberation as factors in the auditory perception of egocentric distance. *Perception & Psychophysics*, 18(6):409–415, 1975.

[194] E. Meyer and G. R. Schodder. Über den Einfluss von Schallrückwürfen auf Richtungslokalisation and Lautstärke bei Sprache [On the influence of reflected sound on directional localization and loudness of speech]. *Nachr. Akad. Wiss. Göttingen*, 6:31–42, 1979. Math. Phys. Klasse IIa.

[195] A. W. Mills. On the minimum audible angle. *J. Acoust. Soc. Am.*, 30:237–246, 1958.

[196] A.W. Mills. Lateralization of high-frequency tones. *J. Acoust. Soc. Am.*, 32:132–134, 1960.

[197] P. J. Minnaar, S. K. Olesen, F. Christensen, and H. Møller. The importance of head movements for binaural room synthesis. In: *Proc. ICAD*, Espoo, Finland, July 2001.

[198] H. Møller, D. Hammershøi, C. B. Jensen, and M. F. Sørensen. Evaluation of artificial heads in listening tests. *J. Audio Eng. Soc.*, 47:83–100, 1999.

[199] H. Møller, M. F. Sørensen, C. B. jensen, and D. Hammershøi. Binaural technique: Do we need individual recordings? *J. Audio Eng. Soc.*, 44:451–469, 2006.

[200] J. K. Moore. The human auditory brain stem: a comparative view. *Hear. Res.*, 29:1–32, 1987.

[201] M. Morimoto and Z. Maekawa. Auditory spaciousness and envelopment. In: *Proc. 13th Int. Congr. on Acoustics*, volume 2, pages 215–218, Belgrade, 1989.

[202] F. Nater, and J. Breebaart. A parametric approach to Head-Related Transfer Functions. Technical Report, École Polytechnique Fédérale de Lausanne, 2006.

[203] T. Okano, L. L. Beranek, and T. Hidaka. Relations among interaural cross-correlation coefficient (IACC$_E$), lateral fraction (LF$_E$), and apparent source width (asw) in concernt halls. *J. Acoust. Soc. Am.*, 104(1):255–265, July 1998.

[204] A. R. Palmer, D. McAlpine, and D. Jiang. Processing of interaural delay in the inferior colliculus. In: J. Syka, editor, *Acoustical signal processing in the central auditory system*, pages 353–364. Plenum Press, New York, 1997.

[205] T. J. Park. IID sensitivity differs between two principal centers in the interaural intensity difference pathway: the LSO and the IC. *J. Neurophysiol.*, 79:2416–2431, 1998.

[206] T. J. Park, P. Monsivais, and G. D. Pollak. Processing of interaural intensity differences in the LSO: role of interaural threshold differences. *J. Neurophysiol.*, 77:2863–2878, 1997.

[207] R. D. Patterson, M. H. Allerhand, and C. Giguère. Time-domain modeling of peripheral auditory processing: A modular architecture and software platform. *J. Acoust. Soc. Am.*, 98(4):1890–1894, October 1995.

[208] S. Perrett and W. Noble. The effect of head rotations on vertical plane sound localization. *J. Acoust. Soc. Am.*, 102:2325–2332, 1997.

[209] D. R. Perrott and A. D. Musicant. Minimum auditory movement angle: binaural localization of moving sound sources. *J. Acoust. Soc. Am.*, 62:1463–1466, 1977.

[210] V. Pulkki. Localization of amplitude-panned sources I: Stereophonic panning. *J. Audio Eng. Soc.*, 49(9):739–752, 2001.

[211] V. Pulkki. Localization of amplitude-panned sources II: Two- and three-dimensional panning. *J. Audio Eng. Soc.*, 49(9):753–757, 2001.

[212] V. Pulkki. Virtual sound source positioning using Vector Base Amplitude Panning. *J. Audio Eng. Soc.*, 45:456–466, June 1997.

[213] M. Purat, T. Liebchen, and P. Noll. Lossless transform coding of audio signals. In: *Preprint 102th Conv. Aud. Eng. Soc.*, Mar. 1997.

[214] H. Purnhagen. Low complexity parametric stereo coding in MPEG-4. In: *Proc. Digital Audio Effects Workshop (DAFX)*, Naples, IT, October 2004. available: http://dafx04.na.infn.it/.

[215] H. Purnhagen and N. Meine. HILN – The MPEG-4 parametric audio coding tools. In: *Proc. ISCAS*, May 2000.

[216] H. Purnhagen, N. Meine, and B. Edler. Sinusoidal coding using loudness-based component selection. In: *Proc. ICASSP*, May 2002.

[217] H. Purnhagen, J. Engdegard, W. Oomen, and E. Schuijers. Combining low complexity parametric stereo with High Efficiency AAC. ISO/IEC JTC1/SC29/WG11 MPEG2003/M10385, December 2003.

[218] H. Purnhagen, J. Engdegard, W. Oomen, and E. Schuijers. Combining low complexity parametric stereo with high efficiency aac. In: *ISO/IEC JTC1/SC29/WG11 N6130*, Dec. 2003.

[219] B. Rakerd and W. M. Hartmann. Localization of sound in rooms, II: The effects of a single reflecting surface. *J. Acoust. Soc. Am.*, 78(2):524–533, Aug. 1985.

[220] B. Rakerd and W. M. Hartmann. Localization of sound in rooms, III: Onset and duration effects. *J. Acoust. Soc. Am.*, 80(6):1695–1706, Dec. 1986.

[221] G. H. Recanzone, S. D. D. R. Makhamra, and D. C. Guard. Comparisons of relative and absolute sound localization ability in humans. *J. Acoust. Soc. Am.*, 103:1085–1097, 1998.

[222] D.E. Robinson and L.A. Jeffress. Effect of varying the interaural noise correlation on the detectability of tonal signals. *J. Acoust. Soc. Am.*, 35:1947–1952, 1963.

[223] T. Robinson. Simple lossless and near-lossless waveform compression. In: *Technical report CUED/F-INFENG/TR.156, Cambridge University, Engineering Department*, Dec. 1994.

[224] J. E. Rose, N. B. Gross, C. D. Geisler, and J. E. Hind. Some neural mechanisms in the inferior colliculus of cat which may be relevant to localization of a sound source. *J. Neurophysiol.*, 29:288–314, 1966.

[225] R.C. Rowland Jr and J.V. Tobias. Interaural intensity difference limen. *J. Speech Hear. Res.*, 10:733–744, 1967.

[226] M. A. Ruggero. *The mammalian auditory pathway: neurophysiology*, chapter 2. Physiology and conding of sound in the auditory nerve. Springer-Verslag, 1992.

[227] F. Rumsey. *Spatial Audio*. Focal Press, Music Technology Series, 2001.

[228] F. Rumsey. Spatial quality evaluation for reproduced sound: Terminology, meaning, and a scene-based paradigm. *J. Audio Eng. Soc.*, 50(9):651–666, 2002.

[229] B. Sayers. Acoustic image lateralization judgments with binaural tones. *J. Acoust. Soc. Am.*, 36:923–926, 1964.

[230] E. D. Scheirer. Structured audio and effects processing in the MPEG-4 multi-media standard. *Multimedia Systems*, 7(1):11–22, 1999.

[231] M. R. Schroeder. An artificial stereophonic effect obtained from a single audio signal. *J. Acoust. Soc. Am.*, 6:74–79, 1958.

[232] M. R. Schroeder. Improved quasi-stereophony and 'colorless' artificial reverberation. *J. Acoust. Soc. Am.*, 33(9):1061–1064, Aug. 1961.

[233] E. Schuijers, W. Oomen, A. C. den Brinker, and A. J. Gerrits. Advances parametric coding for high-quality audio. In: *Proc. MPCA*, Nov. 2002.

[234] E. Schuijers, W. Oomen, B. den Brinker, and J. Breebaart. Advances in parametric coding for high-quality audio. In: *Preprint 5852, 114th AES convention, Amsterdam, The Netherlands*, 2003.

[235] E. Schuijers, J. Breebaart, H. Purnhagen, and J. Engdegård. Low complexity parametric stereo coding. In: *Preprint 5852, Proc. 116th AES convention, Berlin, Germany*, 2004.

[236] B. G. Shinn-Cunningham. The perceptual consequences of creating a realistic, reverberant 3-D audio display. In: *Proc. of the international congress on acoustics*, Kyoto, Japan, April 2004.

[237] B. G. Shinn-Cunningham. Applications of virtual auditory displays. In: *20th Ann. Conf. IEEE Eng. Med. Biol. Soc.*, Hong Kong, China, October 1998.

[238] B. G. Shinn-Cunningham. Distance cues for virtual auditory space. In: *Proc. 1st IEEE Pacific-Rim Conf. on Multimedia, Sydney, Australia*, pages 227–230, Dec. 2000.

[239] B. G. Shinn-Cunningham, P. M. Zurek, and N. I. Durlach. Adjustment and discrimination measurements of the precedence effect. *J. Acoust. Soc. Am.*, 93:2923–2932, 1993.

[240] B. G. Shinn-Cunningham, S. Santarelli, and N. Kopco. Tori of confusion: binaural localization cues for sources within reach of a listener. *J. Acoust. Soc. Am.*, 107:1627–1636, 2000.

[241] P. K. Singh, Y. Ando, and Y. Kurihara. Individual subjective diffuseness responses of filtered noise sound fields. *Acustica*, 80:471–477, July 1994.

[242] S. Singhal. High quality audio coding using multipulse LPC. In: *Proc. Int. Conf. Acoust., Speech, and Signal Proc.*, volume I, pages 1101–1104, Albuquerque, USA, April 1990. IEEE.

[243] D. Sinha, J. D. Johnston, S. Dorward, and S. Quackenbush. The perceptual audio coder (PAC). In V. Madisetti and D. B. Williams, editors, *The Digital Signal Processing Handbook*, chapter 42. CRC Press, IEEE Press, Boca Raton, Florida, 1997.

[244] J. O. Smith and X. Serra. PARSHL: An analysis/synthesis program for nonharmonic sounds based on sinusoidal representation. In: *Proc. Int. Computer Music Conf.*, pages 290–297, 1987.

[245] J. Steinberg and W. Snow. Auditory perspectives – physical factors. *Stereophonic Techniques*, pages 3–7, 1934. Audio Engineering Society.

[246] J. Steinberg and W. Snow. Auditory perspectives – physical factors. *Electrical Engineering*, 53(1):12–15, Jan. 1934.

[247] R. M. Stern and G. D. Shear. Lateralization and detection of low-frequency binaural stimuli: Effects of distribution of internal delay. *J. Acoust. Soc. Am.*, 100:2278–2288, 1996.

[248] R. M. Stern, A. S. Zeiberg, and C. Trahiotis. Lateralization of complex binaural stimuli: A weighted-image model. *J. Acoust. Soc. Am.*, 84:156–165, 1988.

[249] G. Stoll. ISO-MPEG-2 audio: A generic standard for the coding of two-channel and multichannel sound. In N. Gilchrist and C. Grewin, editors, *Collected Papers on Digital Audio Bit-Rate Reduction*, pages 43–53. Audio Engineering Society Inc., 1996.

[250] R. Streicher and F. A. Everest. *The New Stereo Soundbook – Second Edition*. Audio Engineering Associates, Pasadena, CA, 1998.

[251] J. W. Strutt (Lord Rayleigh). On our perception of sound direction. *Philos. Mag.*, 113: 214–232, 1907.

[252] G. Theile. Zur Kompatibilität von Kunstkopfsignalen mit intensitätsstereophonen Signalen bei Lautsprecherwiedergabe: Die Klangfarbe On the compatibility of dummy-head signals with intensity stereophony signals in loudspeaker reproduction: Timbre. *Rundfunktech. Mitt.*, 25:146–154, 1981.

[253] G. Theile. Zur Theorie der optimalen Wiedergabe von stereophonen Signalen über Lautsprecher und Kopfhörer on the theory of the optimal reproduction of stereophonic signals over loudspeakers and headphones. *Rundfunktech. Mitt.*, 25:155–169, 1981.

[254] G. Theile. Zur Kompatibilität von Kunstkopfsignalen mit intensitätsstereophonen Signalen bei Lautsprecherwiedergabe: Die Richtungsabbildung On the compatibility of dummy-head signals with intensity stereophony signals in loudspeaker reproduction: Directional imaging. *Rundfunktech. Mitt.*, 25:67–73, 1981.

[255] G. Theile. On the performance of two-channel and multi-channel stereophony. In: *Preprint 88th Conv. Aud. Eng. Soc.*, March 1990.

[256] G. Theile and G. Plenge. Localization of lateral phantom sources. *J. Audio Eng. Soc.*, 25(4): 196–200, 1977.

[257] C. Tsuchitani. Input from the medial nucleus of trapezoid body to an interaural level detector. *Hear. Res.*, 105:211–224, 1997.

[258] K. Tsutsui, H. Suzuki, O. Shimoyoshi, M. Sonohara, K. Akagiri, and R. M. Heddle. ATRAC: Adaptive transform acoustic coding for MiniDisc. In N. Gilchrist and C. Grewin, editors, *Collected Papers on Digital Audio Bit-Rate Reduction*, pages 95–101. Audio Engineering Society Inc., 1996.

[259] S. van de Par and A. Kohlrausch. A new approach to comparing binaural masking level differences at low and high frequencies. *J. Acoust. Soc. Am.*, 101:1671–1680, 1997.

[260] S. van de Par and A. Kohlrausch. Diotic and dichotic detection using multiplied-noise maskers. *J. Acoust. Soc. Am.*, 103:2100–2110, 1998.

[261] S. van de Par, A. Kohlrausch, J. Breebaart, and M. McKinney. Discrimination of different temporal envelope structures of diotic and dichotic target signals within diotic wide-band noise. In D. Pressnitzer, A. de Cheveigné, S. McAdams, and L. Collet, editors, *Auditory signal processing: physiology, psychoacoustics, and models*, volume Proc. 13th int. symposium on hearing. Springer Verlag, New York, 2004.

[262] S. van de Par, O. Schimmel, A. Kohlrausch, and J. Breebaart. Source segregation based on temporal envelope structure and binaural cues. In B. Kollmeier, G. Klump, V. Hohmann, U. Langemann, M. Mauermann, S. Uppenkamp, and J. Verhey, editors, *Hearing – from research to applications*, Heidelberg, Germany, 2007. Springer Verlag.

[263] M. van der Heijden and C. Trahiotis. Binaural detection as a function of interaural correlation and bandwidth of masking noise: Implications for estimates of spectral resolution. *J. Acoust. Soc. Am.*, 103:1609–1614, 1998.

[264] E. N. G. Verheijen. *Sound Reproduction by Wave Field Synthesis*. PhD thesis, Delft University of Technology, 1997.

[265] L. Villemoes, J. Herre, J. Breebaart, G. Hotho, S. Disch, H. Purnhagen, and K. Kjörling. MPEG Surround: the forthcoming ISO standard for spatial audio coding. In: *Proc. 28th AES conference*, Pitea, Sweden, 2006.

[266] E. M. von Hornbostel and M. Wertheimer. über die Wahrnehmung der Schallrichtung [On the perception of the direction of sound]. *Sitzungsber. Akad. Wiss. Berlin*, pages 388–396, 1920.

[267] R.G.v.d. Waal and R.N.J. Veldhuis. Subband coding of stereophonic digital audio signals. *Proc. IEEE ICASSP 1991*, pages 3601–3604, 1991.

[268] H. Wallach, E. B. Newman, and M. R. Rosenzweig. The precedence effect in sound localization. *AM. J. Psychol.*, 62:315–336, 1949.

[269] T. F. Weis and C. Rose. A comparison of synchronization filters in different auditory receptor organs. *Hear. Res.*, 33:175–180, 1988.

[270] E. M. Wenzel and S. H. Foster. Perceptual consequences of interpolating head-related transfer functions during spatial synthesis. In: *Proceedings of the 1993 workshop on applications of signal processing to audio and acoustics*, New York, 1993.

[271] E. M. Wenzel, M. Arruda, D. J. Kistler, and F. L. Wightman. Localization using nonindividualized head-related transfer functions. *J. Acoust. Soc. Am.*, 94:111–123, 1993.

[272] F. L. Wightman and D. J. Kistler. Headphone simulation of free-field listening. I. Stimulus synthesis. *J. Acoust. Soc. Am.*, 85:858–867, 1989.

[273] F. L. Wightman and D. J. Kistler. Headphone simulation of free-field listening. II: Psychophysical validation. *J. Acoust. Soc. Am.*, 85:868–878, 1989.

[274] F. L. Wightman and D. J. Kistler. The dominant role of low-frequency interaural time differences in sound localization. *J. Acoust. Soc. Am.*, 91:1648–1661, 1992.

[275] F. L. Wightman and D. J. Kistler. Resolution of front-back ambiguity in spatial hearing by listener and source movement. *J. Acoust. Soc. Am.*, 105:2841–2853, 1999.

[276] F. L. Wightman and D. J. Kistler. Individual differences in human sound localization behavior. *J. Acoust. Soc. Am.*, 99:2470–2500, 1996.

[277] R.H. Wilson and C.G. Fowler. Effects of signal duration on the 500-Hz masking-level difference. *Scand. Audio.*, 15:209–215, 1986.

[278] R.H. Wilson and R.A. Fugleberg. Influence of signal duration on the masking-level difference. *J. Speech Hear. Res.*, 30:330–334, 1987.

[279] M. Wolters, K. Kjörling, D. Homm, and H. Purnhagen. A Closer Look into MPEG-4 High Efficiency AAC. In: *Proc. 115th AES Convention*, Los Angeles, USA, October 2003. Preprint 5871.

[280] Tom C. T. Yin and Joseph C. K. Chan. Interaural time sensitivity in medial superior olive of cat. *J. Neurophysiol.*, 64(2):465–488, August 1990.

[281] W. A. Yost. Tone-in-tone masking for three binaural listening conditions. *J. Acoust. Soc. Am.*, 52:1234–1237, 1972.

[282] W. A. Yost. Weber's fraction for the intensity of pure tones presented binaurally. *Percept. Psychophys.*, 11:61–64, 1972.

[283] W. A. Yost. Discrimination of interaural phase differences. *J. Acoust. Soc. Am.*, 55: 1299–1303, 1974.

[284] W. A. Yost. Lateral position of sinusoids presented with interaural intensive and temporal differences. *J. Acoust. Soc. Am.*, 70:397–409, 1981.

[285] W. A. Yost and E. R. Hafter. *Lateralization*. Yost & Gourevich, 1991.

[286] W. A. Yost, D. W. Nielsen, D. C. Tanis, and B. Bergert. Tone-on-tone binaural masking with an antiphasic masker. *Percept. Psychophys.*, 15:233–237, 1974.

[287] R. Zelinski and P. Noll. Adaptive transform coding of speech signals. *IEEE Trans. Acoust. Speech, and Signal Processing*, 25:299–309, August 1977.

[288] M. Zhang, K. Tan, and M. H. Er. Three-dimensional sound synthesis based on head-related transfer functions. *J. Audio. Eng. Soc.*, 146:836–844, 1998.

[289] P. M. Zurek. The precedence effect and its possible role in the avoidance of interaural ambiguities. *J. Acoust. Soc. Am.*, 67:952–964, 1980.

[290] P. M. Zurek. Probability distributions of interaural phase and level differences in binaural detection stimuli. *J. Acoust. Soc. Am.*, 90.1927–1932, 1991.

[291] P. M. Zurek. The precedence effect. In W. A. Yost and G. Gourevitch, editors, *Directional Hearing*, pages 85–105. Springere Verlag, New York, 1987.

[292] P. M. Zurek and N. I. Durlach. Masker-bandwidth dependence in homophasic and antiphasic tone detection. *J. Acoust. Soc. Am.*, 81:459–464, 1987.

[293] E. Zwicker and H. Fastl. *Psychoacoustics: Facts and Models*. Springer, New York, 1999.

[294] U. T. Zwicker and E. Zwicker. Binaural masking-level difference as a function of masker and test-signal duration. *Hearing Research*, 13:215–220, 1984.

[295] J. Zwislocki and R. S. Feldman. Just noticeable differences in dichotic phase. *J. Acoust. Soc. Am.*, 28:860–864, 1956.

Index

Printed and bound by CPI Group (UK) Ltd, Croydon, CR0 4YY

16/04/2025

14658391-0002